Lecture Notes in Mathematics

A collection of informal reports and seminars
Edited by A. Dold, Heidelberg and B. Eckmann, Zürich

T0222664

3

J. Frank Adams

Department of Mathematics, University of Manchester

Stable Homotopy Theory

Third Edition
Lectures delivered at the University of California
at Berkeley 1961. Notes by A. T. Vasquez

Springer-Verlag
Berlin · Heidelberg · New York 1969

National Science Foundation Grant 10 700

All rights reserved. No part of this book may be translated or reproduced in any form without written permission from
Springer Verlag. © by Springer-Verlag Berlin · Heidelberg 1969
Library of Congress Catalog Card Number 70–90867 · Printed in Germany. Title No. 7323

TABLE OF CONTENTS

1) Introduction

Before I get down to the business of exposition, I'd like to offer a little motivation. I want to show that there are one or two places in homotopy theory where we strongly suspect that there is something systematic going on, but where we are not yet sure what the system is.

The first question concerns the stable J-homomorphism. I recall that this is a homomorphism

$$J: \pi_r(SO) \to \pi_r^S = \pi_{r+n}(S^n), \text{ n large.}$$

It is of interest to the differential topologists. Since Bott, we know that $\pi_r(SO)$ is periodic with period 8:

$r =$	1	2	3	4	5	6	7	8	9...
$\pi_r(SO) =$	Z_2	0	Z	0	0	0	Z	Z_2	Z_2...

On the other hand, π_r^S is not known, but we can nevertheless ask about the behavior of J. The differential topologists prove:

<u>Theorem</u>: If $r = 4k - 1$, so that $\pi_r(SO) \cong Z$, then $J(\pi_r(SO)) = Z_m$ where m is a multiple of the denominator of $B_k/4k$ (B_k being in the kth Bernoulli number.)

<u>Conjecture</u>: The above result is best possible, i.e. $J(\pi_r(SO)) = Z_m$ where m is exactly this denominator.

<u>Status of conjecture</u>: No proof in sight.

<u>Conjecture</u>: If $r = 8k$ or $8k + 1$, so that $\pi_r(SO) = Z_2$, then $J(\pi_r(SO)) = Z_2$.

<u>Status of conjecture</u>: Probably provable, but this is work in progress.

The second question is somewhat related to the first; it concerns vector fields on spheres. We know that S^n admits a continuous field of non-zero tangent vectors if and only if n is odd. We also know that if $n = 1,3,7$ then S^n is parallelizable: that is, S^n admits n continuous tangent vector fields which are linearly independent at every point. The question is then: for each n, what is the maximum number, $r(n)$, such that S^n admits $r(n)$ continuous tangent vector fields that are linearly independent at every point? This is a very classical problem in the theory of fibre bundles. The best positive result is due to Hurwitz, Radon and Eckmann who construct a certain number of vector fields by algebraic methods. The number, $\rho(n)$, of fields which they construct is always one of the numbers for which $\pi_r(SO)$ is not zero (0,1,3,7,8,9,11....). To determine which, write $n + 1 = (2t +1)2^{\nu}$: then $\rho(n)$ depends only on ν and increasing ν by one increases $\rho(n)$ to the next allowable value.

<u>Conjecture</u>: This result is best possible: i.e. $\rho(n) = r(n)$.

<u>Status of conjecture</u>: This has been confirmed by Toda for $\nu < 11$.

It seems best to consider separately the cases in which $\rho(n) = 8k - 1, 8k, 8k + 1, 8k + 3$. The most favourable case appears to be that in which $\rho(n) = 8k + 3$. I have a line of investigation which gives hope of proving that the result is best possible in this case.

Now, I. M. James has shown that if S^{q-1} admits r-fields, then S^{2q-1} admits r + 1 fields. Therefore the proposition that $\rho(n) = r(n)$ when $\rho(n) = 8k + 3$ would imply that $r(n) \leq \rho(n) + 1$ in the other three cases. This would seem to show that the result is in sight in these cases also: either one can try to refine the inference based on James' result or one can try to adapt the proof of the case $\rho(n) = 8k + 3$ to the case $\rho(n) = 8k + 1$.

2) Primary operations

It is good general philosophy that if you want to show that a geometrical construction is possible, you go ahead and perform it; but if you want to show that a proposed geometric construction is impossible, you have to find a topological invariant which shows the impossibility. Among topological invariants we meet first the homology and cohomology groups, with their additive and multiplicative structure. Afte that we meet cohomology operations, such as the celebrated Steenrod square. I recall that this is a homomorphism

$$Sq^i: H^n(X,Y;Z_2) \to H^{n+i}(X,Y;Z_2)$$

defined for each pair (X,Y) and for all non-negative integers i and n. (H^n is to be interpreted as singular cohomology.) The Steenrod square enjoys the following properties:

1) Naturality: if $f:(\overline{X},\overline{Y}) \to (X,Y)$ is a map, then $f^*(Sq^i u) = Sq^i f^* u$.
2) Stability: if $\delta: H^n(Y;Z_2) \to H^{n+1}(X,Y;Z_2)$ is the coboundary homomorphism of the pair (X,Y), then $Sq^i(\delta u) = \delta(Sq^i u)$
3) Properties for small values of i.
   i) $Sq^0 u = u$
   ii) $Sq^1 u = \beta u$ where β is the Bockstein coboundary associated with the exact sequence $0 \to Z_2 \to Z_4 \to Z_2 \to 0$.

4) Properties for small values of n.

i) if $n = i$ $Sq^i u = u^2$

ii) if $n < i$ $Sq^i u = 0$.

5) Cartan formula:

$$Sq^i(u \cdot v) = \sum_{j+k=i} (Sq^j u) \cdot (Sq^k v)$$

6) Adam relations: if $i < 2j$ then

$$Sq^i Sq^j = \sum_{\substack{k+\ell = i+j \\ k \geq 2\ell}} \lambda_{k,l}\, Sq^k Sq^\ell$$

where the $\lambda_{k,\ell}$ are certain binomial coefficients which one finds in Adam's paper [1].

References for these properties are found in Serre[2]. These properties are certainly sufficient to characterize the Steenrod squares axiomatically; as a matter of fact, it is sufficient to take fewer properties, namely 1, 2, and 4(i).

Perhaps one word about Steenrod's definition is in order. One begins by recalling that the cup-product of cohomology classes satisfies

) $u \cdot v = (-1)^{pq} v \cdot u$ where

$u \in H^p(X;Z)$ and $v \in H^q(X;Z)$.

However the cup-product of cochains does not satisfy this rule. One way of proving this rule is to construct, more or ess explicitly, a chain homotopy: to every pair of cochains, ., y, one assigns a cochain, usually written $x \smile_1 x$, so that

$$\delta(x \smile_1 y) = xy - (-1)^{pq} yx$$

if x and y are cocycles of dimension p and q respectively. Therefore if x is a mod 2 cocycle of dimension m $\delta(x \smile_1 x) = xx \pm xx = 0$ mod 2. We define $Sq^{n-1}x = \{x \smile_1 x\}$, the mod 2 cohomology class of the cocycle $x \smile_1 x$. Steenrod's definition generalized this procedure.

The notion of a primary operation is a bit more general. Suppose given n,m,G,H where n,m are non-negative integers and G and H are abelian groups. Then a primary operation of type (n,m,G,H) would be a <u>function</u>

$$\phi\colon H^n(X,Y;G) \to H^m(X,Y;H)$$

defined for each pair (X,Y) and natural with respect to mappings of such pairs.

Similarly, we define a stable primary operation of degree i. This is a sequence of functions:

$$\phi_n\colon H^n(X,Y;G) \to H^{n+i}(X,Y;H)$$

defined for each n and each pair (X,Y) so that each function ϕ_n is natural and $\phi_{n+1}\delta = \delta\phi_n$ where δ is the coboundary homomorphism of the pair (X,Y). From what we have assumed it can be shown that each function ϕ_n is necessarily a homomorphism.

Now let's take $G = H = Z_2$. Then the stable primary operations form a set A, which is actually a graded algebra because two such operations can be added or composed in the obvious fashion. One should obviously ask, "What is the structure of A?"

Theorem 1. (Serre) A is generated by the Steenrod squares Sq^i.

(For this reason, A is usually called the Steenrod algebra, and the elements $a \in A$ are called Steenrod operations.)

More precisely, A has a Z_2-basis consisting of the operations

$$Sq^{i_1} Sq^{i_2} \ldots Sq^{i_t}$$

where $i_1, \ldots, i_t$ take all values such that

$$i_r \geq 2i_{r+1} \quad (1 \leq r < t) \quad \text{and} \quad i_t > 0.$$

The empty product is to be admitted and interpreted as the identity operation.

(The restriction $i_r \geq 2i_{r+1}$ is obviously sensible in view of property 6) listed above.) There is an analogous theorem in which Z_2 is replaced by Z_p.

Remark: The products $Sq^{i_1} \ldots Sq^{i_t}$ considered above are called admissible monomials. It is comparatively elementary to show that they are linearly independent operations. For example, take $X = \overset{n}{\underset{1}{X}} RP^\infty$, a Cartesian product of n copies of real (infinite dimensional) projective spaces: let $x_i \in H^1(RP^\infty; Z_2)$ be the generators in the separate factors ($i = 1, \ldots, n$), so that $H^*(X; Z_2)$ is a polynomial algebra generated by $x_1, \ldots, x_n$. Then Serre and Thom have shown that the admissible monomials of a given dimension d take linearly independent values on the class $x = x_1 \cdot x_2 \cdots \cdot x_n \in H^n(X;Z_2)$ if n is sufficiently large compared to d.

The computation of $Sq^{i_1} \cdots \cdot Sq^{i_t}$ on the class x is reduced by the Cartan formula to the computation of other iterated operations on the x_i's themselves. Properties 3(i), 4(i) and (ii) imply that $Sq^0 x_i = x_i$, $Sq^1 x_i = x_i^2$, and $Sq^j x_i = 0$ for $j > 1$. The Cartan formula then allows us to compute iterated operations on the x_i's. The details are omitted.

The substance, then, of Theorem 1 is that the admissible monomials span A. This is proved by using Eilenberg-MacLane spaces.

I recall that a space K is called an Eilenberg-MacLane space of type (π,n)--written $K \in K(\pi,n)$--if and only if

$$\pi_r(K) = \begin{cases} \pi & \text{if } r = n \\ 0 & \text{otherwise.} \end{cases}$$

It follows, by the Hurewicz Isomorphism Theorem (if $n > 1$) that $H_r(K) = 0$ for $r < n$ and $H_n(X) \simeq \pi$. Hence $H^n(K;\pi) \simeq \text{Hom}(\pi,\pi)$, and $H^n(K;\pi)$ contains an element b^n, the fundamental class, corresponding to the identity homomorphism from π to π.

Concerning such spaces K, we have

Lemma 1. Let (X,Y) be "good" pair (e.g. homotopy equivalent to a CW-pair.) Let Map $(X,Y;K,k_0)$ denote the set of homotopy classes of mappings from the pair (X,Y) to the pair (K,k_0), k_0 being a point of K. Then this set is in one-to-one correspondence with $H^n(X,Y;\pi)$. The correspondence is given by assigning to each class, {f}, of maps the element f^*b^n.

This lemma is proved by obstruction theory and is classical, see e. g. [3].

Lemma 2. There is a one to one correspondence between cohomology operations ϕ, as defined above, and elements C^m of $H^m(G,n;H)$. The correspondence is given by $\phi \to \phi(b^n)$. The notation $H^m(G,n;H)$ means the cohomology groups (coefficients H) of an Eilenberg-MacLane space of type (G,n), this depends only on G, n and H. b^n is the fundamental class in $H^n(G,n;G)$.

This lemma follows from the first rather easily for "nice " pairs. But a general pair can be replaced by a C-W pair without affecting the singular cohomology.

There is a similar corollary for stable operations. In order to state it, I need to recall that if $K \in K(G,n)$ then its space of loops, ΩK, is an Eilenberg-MacLane space of type (G,n-1). The suspension σ: $H^m(K) \to (\Omega K)$ is defined as follows:

Let K denote the space of paths in K. Then we have : $(LK,\Omega K) \to (K,pt)$, the map that assigns to each path its endpoint. The map σ is the composition:

$$H^m(K) \leftarrow H^m(K,pt) \xrightarrow{\pi^*} H^m(LK,\Omega K) \xleftarrow{\delta} H^{m-1}(\Omega K).$$

The arrows which point the wrong way are conveniently isomorphisms so can be reversed, the last one, δ, is such because LK is a contractible space.

Lemma 3. There is a 1-1 correspondence between stable primary operations, as considered above, and sequences of elements $e^{n+i} \in H^{n+i}(G,n;H)$ (one for each n) such that $\sigma e^{n+i} = e^{n-1+i}$.

We may rephrase this. For n sufficiently large the groups $H^{n+i}(G,n;H)$ may be identified under the map σ for it is then an isomorphism. Any of these isomorphic groups can be called the "stable Eilenberg-MacLane group of degree i". The lemma then asserts that stable primary operations of degree i correspond one to one with the elements stable of the Eilenberg-MacLane group of degree i. For Theorem 1, then, it remains to calculate these groups in the case $G = H = Z_2$.

Theorem 2. (Serre) $H^*(Z_2,n;Z_2)$ is a polynomial algebra, having as generators the classes

$$Sq^{i_1} Sq^{i_2} \ldots Sq^{i_t} b^n$$

where $i_1, \ldots, i_t$ take all values such that

i) $i_1 \geq 2i_2, \ldots, i_{t-1} \geq 2i_t \qquad i_t > 0$

ii) $i_r < i_{r+1} + \ldots + i_t + n \quad$ for each r.

The empty sequence is again allowed and interpreted as indicating the fundamental class b^n.

Remark: These restrictions are obviously sensible in view of properties 4 and 6 above. The conditions are not all independent but this does not worry us.

The proof of the theorem proceeds by induction on n. We know that $H^*(K(Z_2,1);Z_2)$ is a polynomial algebra on one generator b^1 because RP^∞ qualifies as a $K(Z_2,1)$. The inductive step consists in arguing from $H^*(Z_2,n;Z_2)$ to $H^*(Z_2,n+1;Z_2)$ by applying the little Borel theorem to the fibering $\Omega K \to LK \to K$ mentioned above where $K \in K(Z_2,n+1)$.

Let me recall the little Borel theorem.

Classes $f_1,f_2,\ldots,f_i,\ldots$ in $H^*(F;Z_2)$ are said to form a <u>simple system of generators</u> if and only if the products $f_1^{\epsilon_1} f_2^{\epsilon_2} \ldots f_n^{\epsilon_n}$ ($\epsilon_i = 0$, or 1) form a Z_2-basis for $H^*(F;Z_2)$.

<u>Theorem 3</u>. (Borel) Let $F \to E \to B$ be a fibration with B simply connected and E contractible. Let $b_1,b_2,\ldots$ be classes in $H^*(B;Z_2)$ such that only a finite number of them lie in any one group $H^n(B;Z_2)$ <u>and</u> such that $\{\sigma(b_i)\}$ is a simple system of generators in $H^*(F;Z_2)$. <u>Then</u> $H^*(B;Z_2)$ is a polynomial algebra generated by $b_1,b_2,\ldots$.

For example, in $H^*(Z_2,1;Z_2)$ the classes b^1, $(b^1)^2$, $(b^1)^4$, $(b^1)^8$, ... form a simple system of generators. Also in $H^*(Z_2,2;Z_2)$ we have the classes b^2, Sq^1b^2, $Sq^2Sq^1b^2$, ... and

$$\sigma(b^2) = b^1$$

$$\sigma(Sq^1b^2 = Sq^1\sigma(b^2) = Sq^1(b^1) = (b^1)^2$$

$$\sigma(Sq^2Sq^1b^2) = Sq^2\sigma(Sq^1b^1) = Sq^2(b^1)^2 = (b^1)^4$$

etc.

Hence $H^*(Z_2,2;Z_2)$ is a polynomial algebra generated by

b^2, $Sq^1 b^2$, $Sq^2 Sq^1 b^2, \ldots$. In a similar way, one argues from $K(Z_2, n)$ to $K(Z_2, n+1)$.

The little Borel theorem is most conveniently proved by using the comparison theorem for spectral sequences. In fact, in the situation of the little Borel theorem, we have two spectral sequences: the first is the spectral sequence of the fibering, and the second is our idea of what the first ought to be. We wish to prove these coincide--which is just what the comparison theorem is for.

However, you have to choose your comparison theorem. The version given by John Moore [4] won't do, because in that version, you have to start on the chain level, and here we wish to start with the E_2 terms. The version given by Chris Zeeman [5] will do very nicely. Zeeman's proof, however, can be greatly simplified in the special case when the E_∞ terms are trivial, and this is the case we need (in fact, it's the only case I've ever needed.)

Before stating the comparison theorem, we recall some notation. A spectral sequence contains a collection of groups $E_r^{p,q}$ $\infty \geq r \geq 2$, p,q integers (Ours will satisfy $E_r^{p,q} = 0$ if $p < 0$ or $q < 0$.) It also contains differentials $d_r \colon E_r^{p,q} \to E_r^{p+r,\, q-r+1}$ such that $d_r \circ d_r = 0$ and such that $H(E_r^{**}; d_r) = E_{r+1}^{**}$. A map, f, between one spectral sequence $\{E_r^{p,q}\}$ and another $\{\overline{E}_r^{p,q}\}$ is a collection of homomorphism $f \colon E_r^{p,q} \to \overline{E}_r^{p,q}$ which commute with the d_r's in an obvious wa

<u>Theorem 4</u>. <u>Comparison Theorem for Spectral Sequences</u>. Let f be a map between two spectral sequences $E_r^{p,q}$ and $\overline{E}_r^{p,q}$ such that:

i) If $f: E_2^{p,o} \simeq \overline{E}_2^{p,o}$ for $p \leq P$

Then $f: E_2^{p,q} \simeq \overline{E}_2^{p,q}$ for $p \leq P$, all q

ii) $E_\infty^{p,q} = 0$ $\overline{E}_\infty^{p,q} = 0$ except for $(p,q) = 0,0)$ in which case

$$f: E_\infty^{0,0} \simeq \overline{E}_\infty^{0,0} .$$

<u>Then</u> $f: E_2^{p,o} \cong E_2^{p,o}$ for all p.

<u>Proof</u>: The proof is by induction on p. The result is true for both $p = 0$ and $p = 1$ by assumption because $E_\infty^{0,0} = E_2^{0,0}$ and $E_\infty^{1,0} = E_\infty^{1,0}$, and similarly for $\overline{E}$, and f is an isomorphism on these E_∞ terms. Now assume that $f: E_2^{p,0} \simeq \overline{E}_2^{p,0}$ for $p \leq P$. Recall that $0 \subset B_2^{p,q} \subset Z_2^{p,q} \subset E_2^{p,q}$ where $B_2^{p,q} = \text{Im } d_2$ and $Z_2^{p,q} = \text{Ker } d_2$, and $H_2 = Z_2^{p,q}/B_2^{p,q} = E_3^{p,q}$ (The tedious superscripts will sometimes be omitted in what follows.) Since d_3 is defined on $E_3^{p,q}$, Im d_3 and Ker d_3 give rise to subgroups B_3 and Z_3 such that $0 \subset B_2 \subset B_3 \subset Z_3 \subset Z_2 \subset E_2^{p,q}$. This process continues; in general we have $0 = B_1 \subset B_2 \subset \cdots \subset B_p \subset Z_{q+1} \subset Z_q \cdots \subset Z_2 \subset Z_1 = E_2^{p,q}$. The quotient group Z_{q+1}/B_p is $E_\infty^{p,q}$, hence zero in our case, at least if

$(p,q) \neq (0,0)$. The boundary map d_r give an isomorphism

$$(Z_{r-1}/Z_r)^{p,q} \xrightarrow{\cong} (B_r/B_{r-1})^{p+r,q-r+1}.$$

Lemma 4. Under the isomorphism $f\colon E_2^{p,q} \to \bar{E}_2^{p,q}$)which holds for $p \leq P$) B_r corresponds to $\bar{B}_r$ and Z_r corresponds to $\bar{Z}_r$ for $p + r \leq P$.

Proof: Again by induction. For $r = 1$ our conventions make it trivial. For $r = 2$ it is also clear. The inductive step is made by inspecting the following diagram in which $p \leq P$.

$$\begin{array}{ccccccc}
(Z_{r-1}/B_{r-1})^{p-r,q+r-1} & \cong & E_r^{p-r,q+r-1} & \xrightarrow{d_r} & E_r^{p,q} & \xrightarrow{\text{mono}} & E_2^{p,q}/B_{r-1}^{p,q} \\
\cong\big\downarrow f & & \big\downarrow f & & \big\downarrow f & & f\big\downarrow\cong \\
(\bar{Z}_{r-1}/\bar{B}_{r-1})^{p-r,q+r-1} & \xrightarrow{\cong} & \bar{E}_r^{p-r,q+r-1} & \xrightarrow{d_r} & \bar{E}_r^{p,q} & \xrightarrow{\text{mono}} & \bar{E}_2^{p,q}/\bar{B}_{r-1}^{p,q}
\end{array}$$

Returning to the main line of argument, we now consider the group $E_2^{p,q}$ where $p + q = P$ $q \geq 1$. By the lemma B_p ($= Z_{q+1}$) is preserved by f and so is Z_q. Therefore $(Z_q/Z_{q+1})^{p,q}$ is mapped isomorphically by f. But

$$(Z_q/Z_{q+1})^{p,q} \xrightarrow[d_{q+1}]{\cong} (B_{q+1}/B_q)^{p+1,0} .$$

Therefore $(B_{q+1}/B_q)^{p+1,0}$ is mapped isomorphically by f (for $1 \leq q \leq P$). Now $E_2^{P+1,0}$ has the composition series $0 = B_1 \subset B_2 \subset \ldots \subset B_{p+1} = Z_1 = E_2^{P+1,0}$. We have just shown that all the successive quotients are mapped isomorphically by f, therefore $E_2^{p+1,0}$ is mapped isomorphically by f. This completes the induction and proves the theorem.

We can now give the proof of theorem 3. The E_2 term of the spectral sequence of the fibering $F \to E \to B$ is $E_2^{p,q} = H^p(B) \otimes H^q(F)$ where Z_2 coefficients are understood and $\otimes$ means tensor with respect to Z_2. (By assumption $H^q(F)$ is finitely generated.) In other words, the cup-product $E_2^{p,0} \otimes E_2^{0,q}$ gives an isomorphism. We will now construct another spectral sequence and apply the comparison theorem. The condition on $E_\infty^{p,q}$ is satisfied because E is a contractible space.

We first construct $\overline{E}_r^{p,q}(i)$ as follows. Let the dimension of b_i be t_i. Let $\overline{E}_2^{p,q}(i)$ have a basis consisting of elements $\overline{b}_i^m \in E_2(i)^{mt_i,0}$ $\overline{f}_i\overline{b}_i^m \in \overline{E}_2^{mt_i,t_i-1}(i)$. Define the differential d_r so that $d_r = 0$ for $2 < r < t_i$ and

$$d_r(\overline{f}_i\overline{b}_i^m) = \overline{b}_i^{m+1} \quad \text{for } r = t_i.$$

Then $E_{r+1}^{*,*}(i)$ has a basis consisting of one element $1 = \overline{b}_i^0$. We set $d_r = 0$ for $r > t_i$.

We now define $\overline{E}$ by

$$\overline{E}_r^{**} = \overline{E}_r^{*,*}(1) \otimes \overline{E}^{**}(2) \otimes \cdots\cdots$$

With the understanding that this is to be interpreted as the direct limit of the finite tensor products. We define d_r by the usual formula.

Note that

$$\begin{aligned} H(\overline{E}_r^{**}) &= H(\overline{E}_r^{**}(1) \otimes \cdots) \\ &= H(\overline{E}_r^{**}(1) \otimes \overline{E}_r^{*,*}(2)) \otimes \cdots \\ &= \overline{E}_{r+1}^{*,*}(1) \otimes \overline{E}_{r+1}^{**}(2) \otimes \cdots \\ &= \overline{E}_{r+1}^{*,*} \end{aligned}$$

We will now define a map $\theta_r \colon \overline{E}_r^{*,*} \to E_r^{*,*}$. Because of our assumption about the relation of the classes f_i to the classes b_i, we have

$$d_r f_i = 0 \qquad \text{for } r < t_i$$

$$d_r f_i = \{b_i\} \qquad \text{for } r = t_i$$

We therefore construct θ_r by setting

$$\theta_r(\overline{f}_i^{\epsilon_i}\, \overline{b}_i^{m_i} \otimes \overline{f}_j^{\epsilon_j}\, \overline{b}_j^{m_j} \otimes \cdots \otimes \overline{f}_\kappa^{\epsilon_k}\, b_k^{m_k})$$

$$= f_i^{\epsilon_i}\{b_i\}^{m_i}\, f_j^{\epsilon_j}\{b_j\}^{m_j} \cdots f_k^{\epsilon_k}\{b_k\}^{m_k}$$

It is immediate that the maps θ_r commute with the d_r.

To apply the comparison theorem we need only check that if $\theta_2 \colon \overline{E}_2^{p,0} \to E_2^{p,0}$ is an isomorphim for $p \leq P$ then $\theta_2 \colon \overline{E}_2^{p,q} \to E_2^{p,q}$ is an isomorphism for all q and $p \leq P$. This is immediate from the follwoing:

$$\begin{array}{ccc}
\bar{E}_2^{p,0} \otimes \bar{E}_2^{0,q} & \xrightarrow[\cong]{\text{cup-product}} & \bar{E}_2^{p,q} \\
\cong \theta_2 \downarrow \otimes \downarrow \theta_2 \cong & & \downarrow \theta_2 \\
E_2^{p,0} \otimes E_2^{0,q} & \xrightarrow{\cong} & E_2^{p,q}
\end{array}$$

The comparison theorem then implies that $\theta_2\colon \bar{E}_2^{*,0} \longrightarrow E_2^{*,0} = H^*(B;Z_2)$ is an isomorphism. Therefore $H^*(B;Z_2)$ is a polynomial algebra generated by the b_1. This completes the proof.

<u>Remark</u>: In the above theorem, the coefficients need not be Z_2, an analogous theorem is valid for coefficients in any commutative ring with identity.

I now wish to turn to Milnor's work [6]. Milnor remarks that the Steenrod algebra is in fact a Hopf algebra. I recall that a Hopf algebra is a graded algebra which is provided with a diagonal homomorphism (of algebras) $\psi\colon A \longrightarrow A \otimes A$. In our case the diagonal ψ is going to be defined by the Cartan-formula 5) so that $\psi(Sq^i) = \sum_{j+k=i} Sq^j \otimes Sq^k$. In general, for any element $a \in A$, there is a unique element $\sum_r a_r' \otimes a_r'' \in A \otimes A$ such that $a(u \cdot v) = \sum_r a_r'(u) \cdot a_r''(v)$. We define $\psi(a) = \sum_r a_r' \otimes a_r''$. I'd better add a word about how this is proven. It is pretty clear that there is such a formula when u and v have some fixed dimensions--say p and q, because it is sufficient to examine the case where u and v are the fundamental classes in $K(Z_2,p) \times K(Z_2,q)$. After that, one has to see that the formula is independent of the p and q. We omit the details.

We ought to check that ψ is a homomorphism, but this just amounts to saying that the two ways of computing $(a\cdot b)(u\cdot v)$ are the same. We ought to check that ψ is associative, but this just amounts to saying that the two ways of computing a(uvw) are the same. Similarly ψ has a co-unit. Thus A is a Hopf algebra.

With any Hopf algebra A over a field K, you can associate the vector space dual:

$$A_t^* = \mathrm{Hom}_K(A_t, K).$$

Assuming that A is finitley generated in each dimension, the structure maps

$$A \otimes A \xrightarrow{\varphi} A \quad A \xrightarrow{\psi} A \otimes A$$

(where φ denotes multiplication in the algebra A) transpose to give

$$A^* \otimes A^* \xleftarrow{\varphi^*} A^* \qquad A^* \xleftarrow{\psi^*} A^* \otimes A^*.$$

Hence A^* is a Hopf algebra.

In our case, the Steenrod Algebra A has a commutative diagonal map but a non-commutative product. By passing to the dual we get an algebra A^* with a commutative product, but a non-commutative diagonal map. For many purposes this is a considerable advantage.

Theorem 5 (Milnor) The Hopf algebra A^*, dual to the Steenrod Algebra, is a polynomial algebra on generators ξ_i of dimension $2^i - 1$ $(i = 1,2,\ldots)$. Since the diagonal map φ^* is a homomorphism it is completely specified by giving its values on the generators: These are

$$\varphi^*\xi_1 = \xi_1 \otimes 1 + 1 \otimes \xi_1$$

$$\varphi^*\xi_2 = \xi_2 \otimes 1 + \xi_1^2 \otimes \xi_1 + 1 \otimes \xi_2$$

$$\varphi^*\xi_i = \sum_{j+k=i} \xi_j^{2^k} \otimes \xi_k \quad (\text{where } \xi_0 = 1)$$

One advantage of this theorem is that it completely determines the multiplicative structure of A without imposing any strain on the memory.

Sketch of proof: We first define the elements ξ_i which are linear functions from A to Z_2. Consider again the space RP^∞ and let $x \in H^1(RP^\infty;Z_2)$ be the generator of H^*. I claim that for any $a \in A$ we have $ax = \sum_{i=0}^{\infty} \lambda_i x^{2^i}$, $\lambda_i \in Z_2$. If this is true then $\lambda_i \in Z_2$ is a function of a having values in Z_2 and is clearly linear, so we can define $\xi_i(a) = \lambda_i$.

The simplest proof that ax has this form is as follows. First recall the definition of a primitive element $H^*(RP^\infty;Z_2)$. We have a product may μ: $RP^\infty \times RP^\infty \longrightarrow RP^\infty$. And h is called primitive if $\mu^* h = h \otimes 1 + 1 \otimes h$. Then we easily check the following:

i) x is primitive

ii) If h is primitive, ah is primitive.

iii) The space of primitive elements is spanned by x^{2^i} $i = 0,1,2,\ldots$. Hence ax has the form stated.

We now wish to show that the monomials $\xi_1^{r_1} \xi_2^{r_2} \ldots \xi_n^{r_n}$ $(r_i \geq 0)$ form a vector space basis for A^*. For this purpose we resurrect our old friends $X = \overset{n}{\underset{i=1}{X}} RP^\infty$, our classes x_i corresponding to the generators in H^1 of each factor, and $x = x_1 \ldots x_n \in H^n(X;Z_2)$. We have previously considered the use of the Cartan formulae to compute a(x). The elegant way of writing this result is

$$ax = \Sigma\, (\xi_{i_1}\, \xi_{i_2} \ldots \xi_{i_n})(a)\; x_1^{2^{i_1}}\; x_2^{2^{i_2}} \ldots x_n^{2^{i_n}}$$

We have previously remarked that the admissible monomials a of a fixed degree (small with respect to n) take linearly indpendent values on x. But this means that a is determined by the values $(\xi_{i_1}\, \xi_{i_2} \ldots \xi_{i_n})(a)$. Therefore the monomials $\xi_1^{r_1} \ldots \xi_k^{r_k}$ span A^*. Since there are precisely the right number of these in each dimension they form a basis for A^*.

It remains to establish the formula for the diagonal map φ^*. This amounts to asking:

if $(ab)x = \Sigma\, \lambda_i x^{2^i}$, what is λ_i in terms of a and b?

But
$$(ab)(x) = a(bx)$$
$$= a(\sum_k \xi_k(b) x^{2^k})$$
$$= \sum_k \xi_k(b) a(x^{2^k}).$$

But $a(x^{2^k})$ can be expanded by the universal rule for expanding $a(x_1 x_2 \ldots x_{2^n})$. We obtain

$\sum_k \xi_k(b) (\sum_j \xi_j^{2^k}(a)\ x^{2^{k+j}}) = \sum_{k,j} \xi_k(b) \xi_j^{2^k}(a)\ x^{2^{k+j}}$. Therefore

$(ab) = \sum_{k+j=m} \xi_k(b) \xi_j^{2^k}(a)$ which is equivalent to what we want.

3) Stable Homotopy Theory

There is a number of phenomena in homotopy theory which are independent of the precise dimensions considered, provided that the dimensions are large enough. For example, $\pi_{n+1}(S^n) = Z_2$ for $n > 2$. Such phenomena, in general, are called stable. One can also point to more complicated theorems (e.g. about spectral sequences) such that each clause of the theorem is true for sufficiently large n, but there is no n which makes all the clauses of the theorem true at once. In proving such a theorem, if you don't take care, you rapidly find yourself carrying a large number of explicit conditions $n > N(p, q, r, \ldots)$, which are not only tedious but basically irrelevant. What we want is a standard convention that we are only considering what happens for sufficiently large n. One approach is to work in a suitably constructed category, in which the objects are not spaces but "stable objects" of some sort. For example, the S-theory of Spanier and Whitehead is such a category.

However S-theory is too restrictive in some ways. I'll give an example. For our purposes, it will be quite essential that our category should contain stable Eilenberg-MacLane objects $K(G,\infty)$. But there are no such objects in S-theory. Again: in one of his papers on cobordism, Milnor wishes to consider the "stable Thom complexes" MSO, MSU corresponding to the groups SO and U. These are justifiable objects, but they don't exist inside S-theory.

I want to go ahead and construct a stable category. Now I should warn you that the proper definitions here are still a matter for much pleasurable argumentation among the experts. The debate is between two attitudes, which I'll personify as the tortoise and the hare. The hare is an idealist: his preferred position is one of elegant and all embracing generality. He wants to build a new heaven and a new earth and no half-measures. If he had to construct the real numbers he'd begin by taking <u>all</u> sequences of rationals, and only introduce that tiresome condition about convergence when he was absolutely forced to.

The tortoise, on the other hand, takes a much more restrictive view. He says that his modest aim is to make a cleaner statement of known theorems, and he'd like to put a lot of restrictions on his stable objects so as to be sure that his category has all the good properties he may need. Of course, the tortoise tends to put on more restrictions than are necessary, but the truth is that the restrictions give him confidence.

You can decide which side you're on by contemplating the Spanier-Whitehead dual of an Eilenberg-MacLane object. This is a "complex" with "cells" in all stable dimensions from $-\infty$ to $-n$. According to the hare, Eilenberg-

MacLane objects are good, Spanier-Whitehead duality is good, therefore this is a good object: And if the negative dimensions worry you, he leaves you to decide whether you are a tortoise or a chicken. According to the tortoise, on the other hand, the first theorem in stable homotopy theory is the Hurewicz Isomorphism Theorem, and this object has no dimension at all where that theorem is applicable, and he doesn't mind the hare introducing this object as long as he is allowed to exclude it. Take the nasty thing away!

Now let's see how these attitudes work out in practice... [The hare proceeds by giving constructions which pass from given categories to new, enlarged categories; some remarks on this subject have been removed from these notes.]

Now let's take a more middle-of-the-road line. This time we talk about various sorts of spectra. A spectrum is a sequence of spaces $\{X_n\}$. We think of X_{n+1} as having higher dimension than X_n and wish to have some comparison between X_n and X_{n+1}. The easiest way is to suppose given maps between SX_n and X_{n+1} or between X_n and ΩX_{n+1}. (Here SX = suspension of X; ΩX = loop space of X). This gives four cases.

1) $f_n: SX_n \longrightarrow X_{n+1}$

2) $g_n: X_{n+1} \longrightarrow SX_n$

3) $h_n : X_n \longrightarrow \Omega X_{n+1}$

4) $k_n : \Omega X_{n+1} \longrightarrow X_n$

Maps of type 1) correspond 1-1 to maps of type 3) so approaches 1) and 3) are equivalent and we get three sorts of spectra.

We may wish to deal with spectra which converge, in some sense. The easiest definition is to say that f_n is an equivalence up to dimension $n + v_n$ where v_n tends to infinity with n. This has analogues in the other sorts of spectra; but if the maps are equivalences, their direction is immaterial.

However, this definition has disadvantages. For example, suppose that we have a sequence of spectra $X^m = \{X_n^m\}$. Try to form the one-point union $\bigvee_m X^m = \{ \bigvee_m V_n^m \}$. Then we have to deal with maps $\bigvee_m f_n^m$ which are equivalences up to dimension $n + \operatorname{Min}_m (v_n^m)$; but $\min_m (v_n^m)$ need not tend to infinity with n.

Now, actually, we have to deal with such constructions. The obvious escape is to specify how fast v_n should tend to infinity. This leads to my chosen definition which is pretty far toward the tortoise end of the scale. It's modeled on J. H. C. Whitehead's idea of building up a complex by attaching cells.

I define a <u>stable complex</u>, X, to be a sequence of CW-complexes X_n which have the following properties:

i) X_n has one vertex and has other r-cells only for $n \leq r \leq 2n - 2$.

ii) The $2n - 3$ skeleton $(X_n)^{2n-3}$ is the reduced suspension $X_{n-1} \# S^1$.

[Here $X \# Y$ means the "smashed" product $= X \times Y / X \vee Y$ where $X \vee Y$ denotes the one point union of two spaces joined at their base points.]

A <u>map</u> $f: X \longrightarrow Y$ between two such stable objects is a sequence of maps:

$$f_n: X_n \longrightarrow Y_n \qquad \text{such that}$$

$$f_n | X_{n-1} \# S^1 = f_{n-1} \# 1.$$

We can compose maps in the obvious fashion.

A <u>homotopy</u> $h: f \sim g$ between two such maps is a sequence of homotopies

$$h_n: (I \times X_n)^{2n-2} \longrightarrow Y_n \quad \text{keeping}$$

base points fixed and commuting with $\# S^1$ in the obvious fashion.

This is equivalent to defining homotopy in terms of an object "$I \times X$" defined as follows:

$$(I \times X_n)_n = \left({}^{I \times X_n}/_{I \times x_0}\right)^{2n-2} \qquad x_0 = \text{vertex of } X_n$$

Whenever I want to apply notions from the general theory of categories, the word morphism is to be interpreted as a homotopy class of mappings. But we allow ourselves to keep the notion of maps so that we may speak of inclusion maps, etc.

Example of a stable object. The stable sphere of dimension r.

We have $X_n = S^{n+r}$ for $n \geq r + 2$

X_n = pt. otherwise.

We have to assume that $r \geq 0$.

Warning. Since spheres of positive dimension only are available in this category it is not always possible to desuspend an object. This is a great blemish from the hare's point of view.

With this category I wish to do three things.

1) To justify it by showing that at least some phenomena of classical stable homotopy theory go over into this category.

2) To make it familiar, by showing that some of the familiar theorems for spaces go over into this category.

3) To lay the foundations for the next lecture by obtaining those properties of the category which I require.

We wish to show that this category does allow us to consider some of the phenomena which are considered in classical stable homotopy theory.

Theorem 1. If K,L are CW complexes with one vertex and positive dimensional cells for $n \leq r \leq 2n - 2$ then there exist stable objects X, Y such that X_n is one of the same homotopy type as K and $X_{r+1} = X_r \# S^1$ for $r \geq n$, and similarly for Y and L. Furthermore, if X, Y have these properties, then Map(X, Y) is in one-one correspondence with Map (K, L).

This follows from the classical suspension theorems, and I wish to say no more about it.

[The notes for the remainder of this lecture have been revised in order to reorganize the proofs.]

Both in stable and in unstable homotopy theory we may take the maps $f: X \to Y$, divide them into homotopy classes, and so form a set Map (X,Y). This set we make into a group (in favourable cases), and such groups figure in certain exact sequences. It is here that a certain basic difference between stable and unstable homotopy theory arises. In unstable homotopy theory we take groups Map(X,Y), and first we try to make exact sequences by varying X . What we need is a pair (X_1, X_2)--that is, an inclusion map with the homotopy extension property. Secondly, we try to make exact sequences by varying Y . In this case we need a fibering, that is, a projection map with the homotopy lifting property. In cohomology the pair gives an exact sequence; the fibering gives a spectral sequence. In stable homotopy theory the distinction disappears: we have just one sort of exact sequence of spaces. In order to construct such exact sequences, suppose given a map $f: X \to Y$ between stable complexes. Then we can construct a new stable complex $Y \cup_f CX$. (Here CX is intended to suggest "the cone on X", and $Y \cup_f CX$ is intended to suggest " Y with the cone CX attached to Y by means of the map f.") The definition is

$$(Y \cup_f CX)_n = (Y_n \cup_{f_n} I \# X_n)^{2n-2} .$$

(Here $I = [0,1]$, with basepoint 0 .)

As indicated above, this construction gives rise to two exact sequences. We will prove this below, but we have first to consider the special case in which Y is a "point" p (that is, Y_n is a point for all n) and f is the "constant map" y . We write SX for the resulting stable complex $P \cup_y CX$, and regard this as the "suspension" in our category. It is clear that a map $f: X_1 \to X_2$ induces a map $Sf: SX_1 \to SX_2$, and similarly for homotopies.

Lemma 1. S: Map $(X,Y) \to$ Map (SX,SY) is a one-to-one correspondence.

Remark 1. The hare would always arrange matters so that this lemma would be a triviality. With the present details, it seems to need proof.

Remark 2. Our proof will involve desuspension. Suppose given a CW-pair K,L of dimension at most $(2n-2)$, and a space Y which is (n-1)-connected; suppose given a map $f: S^1 \# K \to S^1 \# Y$, and a deformation h of $f|S^1 \# L$ into a suspended map $1 \# g$ (for some $g: L \to Y$.) Then we can extend the deformation h and the desuspension g over K. In fact, the map f is equivalent to a map $\tilde{f}: K \to \Omega(S^1 \# Y)$; we are given a deformation $\tilde{h}$ of $\tilde{f}|L$ into a map $g: L \to Y \subset \Omega(S^1 \# Y)$, and we are asked to extend $\tilde{h}, g$ over K; this is trivially possible, since $\pi_r(\Omega(S^1 \# Y), Y) = 0$ for $r < 2n$.

Similar remarks apply, when we are given a map $f: S^1 \# K \# S^1 \to S^1 \# Y \# S^1$, and asked to deform it into a suspension $1 \# g \# 1$, or to extend a deformation already given on $S^1 \# L \# S^1$. One has only to replace $\Omega(S^1 \# Y)$ by $\Omega^2(S^1 \# Y \# S^1)$.

<u>Remark 3</u>. The effect of the definition of SX is that $(SX)_n = (S^1 \# X_n)^{2n-2} = S^1 \# X_n^{2n-3} = S^1 \# X_{n-1} \# S^1$.

<u>Remark 4</u>. We shall deduce lemma 1 from the following lemma.

<u>Lemma 2.</u> Suppose given a pair K, L of stable complexes, a map $f: SK \to SY$ (consisting of $f_n: (SK)_n \to (SY)_n$), a map $g: L \to Y$ (consisting of $g_{n-1}: L_{n-1} \to Y_{n-1}$) and a sequence $h = \{h_n\}$ of homotopies

$$h_{n-1}: 1 \# g_{n-1} \# 1 \sim f_n | S^1 \# L_{n-1} \# S^1 : S^1 \# L_{n-1} \# S^1 \to S^1 \# Y_{n-1} \# S^1$$

such that the h_n commute with $\# S^1$. Then the maps g and h can be extended from L to K so as to preserve these properties.

Proof. We proceed by induction over n. Suppose that the maps g_r and h_r have been extended over K for $r \leq n-1$. We are given $f_{n+1}: S^1 \# K_n \# S^1 \to S^1 \# Y_n \# S^1$, and we wish to construct a certain deformation of it into a map $1 \# g_n \# 1$. By the data, we are given the deformation over L_n. We are also given the deformation over $K_{n-1} \# S^1 \subset K_n$ (by applying $\# S^1$ to h_{n-1}.) We thus obtain a deformation compatibly defined over $L_n \cup (K_{n-1} \# S^1)$. By remark 2, the deformation can be extended over K_n. This completes the induction.

We now turn to the proof of lemma 1. We first prove that S: Map (X,Y) -> Map (SX,SY) is an epimorphism. Suppose given a map f: SX -> SY ; we apply lemma 2, taking K to be X, L to be a "point", g and h trivial. The lemma provides a map g: X -> Y such that $Sg \sim f$.

Secondly, it is necessary to prove that

S: Map (X,Y) -> Map (SX,SY)

is monomorphic. Suppose given two maps f_1, f_2: X -> Y such that $Sf_1 \sim Sf_2$; we will apply lemma 2 again. We take K to be the stable complex "I x X", and L to be its two end; we define g: L -> Y using f_1 and f_2, and we define h: SK -> SY by using the homotopy $Sf_1 \sim Sf_2$. The homotopies h_n can be taken stationary on L. Lemma 2 provides an extension of g over "I x X", that is, a homotopy $f_1 \sim f_2$ in our category. This completes the proof of lemma 1.

For $r \geq 2$ the sets

$$\mathrm{Map}\,(S^r X, S^r Y)$$

form abelian groups, and the product is independent of which "suspension coordinate" in $S^r X$ is used to define the product. (In fact, the arguments which one usually uses for spaces apply, because the constructions involve suspension coordinates "on the left," and commute with the operation $\# S^1$ "on the right" used in defining our category.

It is illuminating to recall that in a category where direct sums and direct products always exist and coincide, the "sum of two morphisms" can be defined purely in terms of composition. (Given two morphisms $f,g\colon X \to Y$, one considers

$$X \xrightarrow{f \times g} Y \times Y \longrightarrow Y \vee Y \xrightarrow{1 \vee 1} Y\,.)$$

In our category direct sums and direct products do exist and coincide; given two stable complexes X and Y, one can define $X \vee Y$ by

$$(X \vee Y)_n = X_n \vee Y_n\,.$$

This is a stable complex which fulfils the axioms both for a direct sum $X \vee Y$ and a direct product $X \times Y$ (at least, as soon as it is equipped with the obvious structure maps.)

The proof that $X \vee Y$ is a direct product may be performed by imbedding $X_n \vee Y_n$ in $X_n \times Y_n$. The desired constructions can easily be performed in $X_n \times Y_n$, and then deformed into $X_n \vee Y_n$, since $\pi_r(X_n \times Y_n, X_n \vee Y_n) = 0$ for $r < 2n$. Details are omitted.

To sum up, we have made our category into an abelian category, and we are entitled to use the following definition.

Definition:

$$\pi_r(X,Y) = \text{Map}\ (S^{n+r}X, S^nY)$$

for any n such that $n \geq 0$, $n + r \geq 0$. (It is not assumed that $r \geq 0$.)

We will now proceed to obtain the two exact sequences mentioned above. We recall that given any map $f\colon X \to Y$ we constructed a stable complex $Y \cup_f CX$; we now write $M = Y \cup_f CX$; we have obvious maps

$$Y \xrightarrow{j} M \xrightarrow{q} SX .$$

Theorem 2. The sequences

$$1) \quad \ldots\pi_r(X,U) \xleftarrow{f^*} \pi_r(X,U) \xleftarrow{j^*} \pi_r(M,U) \xleftarrow{q^*} \pi_{r+1}(X,U) \longleftarrow \ldots$$

ii) $\ldots\pi_r(V,X) \xrightarrow{f_*} \pi_r(V,Y) \xrightarrow{j_*} \pi_r(V,M) \xrightarrow{q_*} \pi_{r-1}(V,X) \to \ldots$

are exact.

To prove this, we follow Puppe's method.

Lemma 3. The sequence

$$\mathrm{Map}\ (X,U) \xleftarrow{f^*} \mathrm{Map}\ (Y,U) \xleftarrow{j^*} \mathrm{Map}\ (M,U)$$

is exact.

The proof is trivial.

Lemma 4. The sequence

(a) $X \xrightarrow{f} Y \xrightarrow{j} (Y \cup_f CX) \xrightarrow{j_1} (Y \cup_f CX) \cup_j CY \xrightarrow{j_2} \ldots$

is equivalent (up to signs) to the sequence

(b) $X \xrightarrow{f} Y \xrightarrow{j} M \xrightarrow{q} SX \xrightarrow{Sf} SY \xrightarrow{Sj} SM \longrightarrow \ldots$

(In sequence (a) each term is constructed from the previous two terms as $Y \cup_f CX$ is constructed from $X \xrightarrow{f} Y$.)

It is clear that lemmas 3 and 4 suffice to prove the exactness of the sequence (i) in theorem 2: moreover, lemma 4 has a similar application to sequence (ii) in theorem 2.

The proof of lemma 4 is unaltered from the usual case. It is sufficient to consider the first four maps in the sequence. The required constructions can be performed "on the left" and commute with the operation $\# S^1$ used in defining our category.

<u>Lemma 5.</u> The sequence

$$\text{Map }(SW,M) \xrightarrow{q_*} \text{Map }(SW,SX) \xrightarrow{(Sf)_*} \text{Map }(SW,SY)$$

is exact.

This lemma show that if we map SW into the sequence (a) of lemma 4, we have exactness at the fourth term $(Y \cup_f CX) \cup_j CY$. Since all subsequent terms are also "fourth terms," we have exactness at all subsequent terms also. This still assumes that the "test-space" SW is suspended at least once; but even so, it proves part (ii) of theorem 2.

<u>Proof of lemma 5.</u> Suppose given a map $\lambda \in \text{Map }(SW,SX)$ such that $(Sf)_*\lambda = 0$. By lemma 1, we can take a representative map for λ which is of the form Sg, where $g \colon W \longrightarrow X$ is a map such that $fg \sim 0$; that is,

$W \xrightarrow{g} X \xrightarrow{f} Y$ can be factored through $W \xrightarrow{i} CW \xrightarrow{h} Y$,

where CW is the "cone on W" and h is a "homotopy." We can now construct a map $m \colon SW \longrightarrow M = Y \cup_f CX$ such that $qm \sim Sg$. In fact, we decompose SW into two "cones" C_+W = "$[0,\frac{1}{2}] \# W$" and C_-W. On C_+W we define m by taking $Cg\colon C_+W \longrightarrow CX$; on C_-W we define m to be h . This completes the proof.

In this category we have homology and cohomology theories defined in the obvious way. We have

$$\cdots \longrightarrow H_{m+n}(X_n) \xrightarrow{\sigma} H_{m+n+1}(X_n \quad S^1) \xrightarrow{i_*} H_{m+n+1}(X_{n+1}) \longrightarrow \cdots ;$$

these groups all become equal after a while, and we define the limit to be $H_m(X)$, where $X = \{X_n\}$. Similarly for the homology maps induced by morphisms. We can define the boundary maps for a pair because our version of suspension has been chosen to commute with ∂ . Similar remarks apply to cohomology.

We now turn to the question of Eilenberg-MacLane objects in this category.

<u>Theorem 3.</u> Suppose that $F = \sum_{t \geq 0} F_t$ is a free graded module over the Steenrod algebra A with finitely many generators in each dimension. Then there exists stable complex K such that:

(i) $H^*(K;Z_2) \cong F$ as an A-module, and

(ii) $\pi_r(X,K) \cong \mathrm{Hom}_A^r(F, H^*(X;Z_2))$

for each stable complex X .

(The symbol Hom_A^r denotes the set of A-maps that lower the degree by r . It is understood that the isomorphism in (ii) is induced by assigning to the map

$f: S^{n+r}X \longrightarrow S^nK$ the associated cohomology homomorphism, composed with appropriate suspension isomorphism.)

Proof. We can construct a stable Eilenberg-MacLane complex of type (Z_2, n) by following J. H. C. Whitehead's procedure, attaching "stable cells" to kill "stable groups". By forming the one point union $K = \bigvee_i K_i$ of such objects, we can arrange it that $H^*(K;Z_2)$ is the required free A-module. The clause about $\pi_r(X,K)$ is just the re-expression in a new guise of lemma 1 of the last lecture--or of the corresponding assertion for maps into a Cartesian proudct of Eilenberg-MacLane spaces.

4) Applications of Homological Algebra to Stable Homotopy Theory

I ought to begin by running through the basic notions of homological algebra in the case where we have graded modules over a graded algebra A. Let M be such a module, i.e. $M = \sum_{0 \leq t \leq \infty} M_t$. $A_s \cdot M_t \subseteq M_{s+t}$. M_t is finitely generated. A resolution of M is a chain complex

$$C_0 \xleftarrow{d} C_1 \leftarrow C_2 \leftarrow \cdots \leftarrow C_s \leftarrow$$

in which i) each C_s is a free graded module over A

ii) each d is an A-map preserving gradation

iii) $H_s(C) = \begin{cases} M & \text{if } s = 0 \\ 0 & \text{if } s > 0 \end{cases}$

This amounts to the same thing as requiring a map $\epsilon: C_0 \to M$ so that

$$0 \leftarrow M \xleftarrow{\epsilon} C_0 \xleftarrow{d} C_1 \leftarrow \cdots \leftarrow C_s \leftarrow \cdots$$

is exact at every stage. Such chain complexes always exist and they are unique up to chain equivalence.

Remark: For the case in which we are interested A = mod 2 Steenrod algebra. Also, in this case, there is no distinction between "free A modules" and "projective A modules".

Let N be another such module: Then define

$$\mathrm{Hom}_A^t(C_s, N) = \text{those (graded) A-maps } f: C_s \to N$$

which lower degree by t, i.e. $f(C_s)_u \subset N_{u-t}$. We have, as usual, maps induced by d,

$$\mathrm{Hom}_A^t(C_o, N) \xrightarrow{d^*} \mathrm{Hom}_A^t(C_1, N) \xrightarrow{d^*} \cdots \to \mathrm{Hom}_A^t(C_s, N) \to \cdots$$

We define $\mathrm{Ext}_A^{s,t}(M, N) = \mathrm{Ker}\, d^*/\mathrm{Im}\, d^*$ in $\mathrm{Hom}_A^t(C_s, N)$. The notation is justified since any two resolutions of M are chain equivalent, and therefore the groups $\mathrm{Ext}_A^{s,t}(M, N)$ depend only on the objects displayed.

Next we go back to stable homotopy theory. We will write $H^*(X)$ for $H^*(X, Z_2)$ to abbreviate notation. We recall that any map $f: X \to Y$ induces an A-map $f^*: H^*(Y) \to H^*(X)$. So we get a function: $\mathrm{Map}\,(X, Y) \to \mathrm{Hom}_A^o(H^*(Y), H^*(X))$. This function is moreover a homomorphism. It is in general neither a monomorphism nor an epimorphism. In order to compute $\mathrm{Map}\,(X, Y)$ by homological methods we need further terms. A general formulation is the following.

Theorem 1. Under suitable conditions on X, Y there exists a spectral sequence whose $E_2^{s,t}$ term is $\mathrm{Ext}_A^{s,t}(H^*(Y), H^*(X))$ and which converges to ${}_2\pi_r(X, Y)$.

The details are as follows:

i) If G is an abelian group, then ${}_2G$ means the quotient of G by the subgroup of elements of odd order. It is clear that elements of odd order play no part in our studies which are confined to the prime 2.

ii) The elements in $E_\infty^{s,t}$ which give a composition series for ${}_2\pi_r(X, Y)$ are those for which $t - s = r$. That is, the "total degree" in the spectral sequence is $t-s$. The filtration of ${}_2\pi_r(X, Y)$ is a decreasing one, so, for example,

$$ {}_2\pi_r(X, Y)/F_1 \cong E_\infty^{o,r} $$

$$ F_1/F_2 \cong E_\infty^{1,r+1} \qquad \text{etc.} $$

iii) The differential d_r raises s by r and decreas $t-s$ by 1.

It is perhaps desirable to give an example of this spectral sequence. Let us take $X = Y = S^o$. The E_2^{**} term is then $\mathrm{Ext}_A^{s,t}(Z_2, Z_2)$ (which is rechristened $H^{s,t}(A)$ for brevity). I recall that $H^{1,*}(A)$ can be identified with the space of primitive elements in A^*. The primitive element $\xi_1^{2^i}$ thus gives us a generator h_i. In $H^{**}(A)$ we can define cup products and this allows us to write down part of a basis for $H^{s,t}(A)$.

s ↑ \ t-s →	0	1	2	3	4	5	6	7
4	h_o^4							$h_o^3 h_3$
3	h_o^3			$h_1^3 = h_o^2 h_2$				$h_o^2 h_3$
2	h_o^2		h_1^2	$h_o h_2$			h_2^2	$h_o h_3$
1	h_o	h_1		h_2				h_3
0	1							

blanks are to be interpreted as 0 groups for $s \leq 4$, $t-s \leq 7$

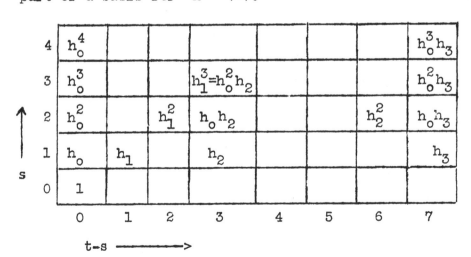

The differentials in this part of the table are all zero, yielding a result in good agreement with the known values of ${}_2\pi_r(\mathcal{S}^0, S^0)$:

r =	0	1	2	3	4	5	6	7
	Z	Z_2	Z_2	Z_8	0	0	Z_2	Z_{16}

Returning to the theorem, we must state suitable conditions on X and Y. We may distinguish two halves to our work:

a) Setting up the spectral sequence. This is more or less formal. I shall assume that $H^*(Y)$ is finitely generated in each dimension because I'll have to assume it later anyway. It is possible, however, that we might be able to eliminate this restriction for this part of the work.

b) Proving the convergence of the spectral sequence.

Existing proofs require the following conditions.

(I) X, the object mapped, must be finitely dimensional, say $X_{r+1} = SX_r$ for $r \geq N$, some N.

(II) $H^*(Y)$ must be finitely generated in each dimension, as assumed above.

It is, perhaps, an interesting exercise for the experts to try to reformulate the theorem so as to relax these conditions. Two changes are fairly obvious. You can replace $\mathrm{Ext}_A^{s,t}(H^*(Y), H^*(X))$ by $\mathrm{Ext}_A^{s,t}(H_*(X), H_*(Y))$ which I believe behaves better with respect to limits; and you can redefine ${}_2G$ by replacing "subgroup of elements of odd order" by "subgroup of elements divisible by arbitrarily high powers of 2." These changes however do not suffice to overcome certain obvious counterexamples. (For example, suppose Y has only one integral homology group which is the group of rationals mod 1.) I have no idea what happens if you replace the coefficient field Z_2 (or Z_p) by the integers or the reals mod 1. (The case Z_p is analogous to the case Z_2.)

Just for variety, however, I want to give a simple and explicit proof of convergence, which works under conditions even more restrictive than I have already stated. That is, I shall assume:

(III) $H^*(Y)$ is a free module over the exterior algebra E generated by Sq^1. This is equivalent to supposing that $H^*(Y; Z)$ has no elements of ∞ order, and all its elements of order 2^f are actually of order 2.

This evidently excludes the case $X = Y = S^0$, so I must give one or two examples to show that it does not exclude all cases of interest.

Ex. 1. Y is the stable object corresponding to RP^{2t}/RP^{2u}. This example is relevant to the vector field problem.

Ex. 2. Set up an exact sequence

$$S^0 \xrightarrow{f} K(Z, 0) \longrightarrow M = K(Z, 0) \cup_f C\, S^0$$

so that $\pi_r(S^0, S^0) \cong \pi_{r+1}(S^0, M) \qquad r > 0$

and $\mathrm{Ext}_A^{s,t}(Z_2, Z_2) \cong \mathrm{Ext}_A^{s-1,t}(H^*(M), Z_2)$

for $t - s > 0$ and $s > 0$.

Condition III is satisfied by M.

Setting up the spectral sequence

Suppose given an object Y and a sequence of order 2,

$$H^*(Y) \xleftarrow{\epsilon} C_0 \xleftarrow{d} C_1 \longleftarrow \cdots \longleftarrow C_s \longleftarrow \cdots \quad \text{where}$$

the C_s's are free modules over A. I don't yet need to suppose that it is a resolution.

By a realization of this sequence, I mean

1) a sequence of Eilenberg-MacLane objects K_s such that $H^*(K_s) \cong C_s$

2) a sequence of objects M_s such that $M_{-1} = Y$

3) maps $f_s\colon M_{s-1} \longrightarrow K_s$ such that $M_s = K_s \cup_{f_s} CM_{s-1}$ and $K_s \xrightarrow{j_s} M_s \xrightarrow{f_{s+1}} K_{s+1}$ induces the map $d\colon C_{s+1} \longrightarrow C_s$, while $f_0\colon M_{-1} \longrightarrow K_0$ induces the map $\epsilon\colon C_0 \longrightarrow H^*(Y)$.

In general, I don't assert that a sequence of order 2 has a realization: but if the sequence is a resolution, then it does. Viz.

We can choose an Eilenberg-MacLane object K_0 such that $H^*(K_0) = C_0$ and a map $f_0\colon Y \longrightarrow K_0$ inducing $\epsilon\colon C_0 \longrightarrow H^*(Y)$. This follows from the last theorem of last lecture. We form the "quotient" $M_o = K_o \cup_{f_o} CY$. We look at the exact sequence

$$\longleftarrow H^t(Y) \xleftarrow[f_o^*]{} H^t(K) \xleftarrow[\rho_o^*]{} H^t(M_o) \longleftarrow \quad .$$

Since f_o^* coincides with ϵ which is onto, ρ_o^* is a monomorphism and $H^*(M_o)$ coincides with $Z_o = \text{Ker}\ \epsilon$. Since $C_1 \xrightarrow{d} C_o \xrightarrow{\epsilon} H^*(Y) \longrightarrow 0$ is exact we have a map $d: C_1 \longrightarrow \text{Ker}\ \epsilon = H^*(M_o)$. We can find K_1, an Eilenberg-MacLane object such that $H^*(K_1) = C_1$ and a map $f_1: M_o \to K_1$ such that $f_1^* = d$. As above, if we form $M_1 = K_1 \cup_{f_1} CM_o$, $H^*(M_1) = Z_1 = \text{Ker}\ d|C_1$. We continue, by induction, forming a sequence of spaces M_s and maps $f_s: M_{s-1} \longrightarrow K_s$ so that $M_s = K_s \cup_{f_s} C\, M_{s-1}$ and

$$M_{s-1} \xrightarrow{f_s} K_s \xrightarrow{\rho_s} M_s \qquad \text{realizes}$$

$$0 \longleftarrow Z_{s-1} \xleftarrow{d} C_s \longleftarrow Z_s \longleftarrow 0 \quad .$$

Given a realization, we can apply the functor $\pi_t(X; \cdot)$ we obtain exact sequences

$$\to \pi_t(X, M_{s-1}) \xrightarrow{f_{s*}} \pi_t(X, K_s) \xrightarrow{\rho_{s*}} \pi_t(X, M_s) \xrightarrow{q_{s*}} \pi_{t-1}(X, M$$

This gives an exact triangle:

$$\begin{array}{ccc} \sum_{s,t} \pi_t(X, M_s) & \xrightarrow{\Sigma q_{s*}} & \sum_{s,t} \pi_t(X, M_{s-1}) \\ & {\scriptstyle \Sigma \rho_{s*}} \nwarrow \quad \swarrow {\scriptstyle \Sigma f_{s*}} & \\ & \sum_{s,t} \pi_t(X, K_s) & \end{array}$$

(In order to avoid trouble it is convenient to get our sequences exact for all s, even $s < 0$. This is done by making a convention that

$$\pi_t(X, K_s) = 0 \quad \text{for} \quad < 0, \quad \pi_t(X; M_s) \xrightarrow[\cong]{q_{s*}} \pi_{t-1}(X; M_{s-1})$$

for $s < 0$.)

Next I remark that this construction is natural. Suppose I have a map $m: Y \longrightarrow \overline{Y}$ inducing $m^*: H^*(Y) \longleftarrow H^*(\overline{Y})$ and that I have two sequences of order two connected by a ladder of maps, as follows:

$$\begin{array}{ccccccccc} H^*(Y) & \longleftarrow & C_0 & \longleftarrow & C_1 & \longleftarrow & C_2 & \longleftarrow & \cdots \\ \uparrow {\scriptstyle m^*} & & \uparrow {\scriptstyle \lambda_0} & & \uparrow {\scriptstyle \lambda_1} & & \uparrow {\scriptstyle \lambda_2} & & \\ H^*(\overline{Y}) & \longleftarrow & \overline{C}_0 & \longleftarrow & \overline{C}_1 & \longleftarrow & \overline{C}_2 & \longleftarrow & \cdots \end{array}$$

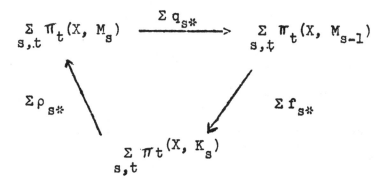

(Such a ladder will always exist if $\{C_s\}$ is a resolution.) We will define the notion of a realization of such a ladder. This consists of a sequence of maps $g_s: K_s \longrightarrow \overline{K}_s$ and a sequence of maps $m_s: M_s \longrightarrow \overline{M}_s$ with the following properties.

1) $g_s^* = \lambda_s$ (for each s.)

2) $m_{-1} = m: Y \longrightarrow \overline{Y}$.

3) For each s, the following diagram is homotopy commutative.

$$\begin{array}{ccccccc}
M_{s-1} & \xrightarrow{f_s} & K_s & \xrightarrow{j_s} & M_s & \xrightarrow{q_s} & SM_{s-1} \\
\downarrow{\scriptstyle m_{s-1}} & & \downarrow{\scriptstyle g_s} & & \downarrow{\scriptstyle m_s} & & \downarrow{\scriptstyle Sm_{s-1}} \\
\overline{M}_{s-1} & \xrightarrow{\overline{f}_s} & \overline{K}_s & \xrightarrow{\overline{j}_s} & \overline{M}_s & \xrightarrow{\overline{q}_s} & S\overline{M}_{s-1}
\end{array}$$

If we have such a realization of a ladder, then we shall obtain induced maps of all our exact sequences, and hence a map of sp ctral sequences. On E_1 the map is induced by the maps λ_s . If we assume that we started with two resolutions, then the induced map on E_2 is given by the induced map of Ext. In particular, if we take $Y = \overline{Y}$, we see that our spectral sequence is defined up to a canonical isomorphism.

<u>Lemma 1</u>. If $\{C_s\}$ is a resolution, then such a ladder can be realized.

The proof is by induction over s. Suppose given $m_{s-1}: M_{s-1} \longrightarrow \overline{M}_{s-1}$. Since $\overline{K}_s$ is an Eilenberg-MacLane object we can construct $g_s: K_s \longrightarrow \overline{K}_s$ such that $\lambda_s^* = g_s$.

We now wish to show that $g_s f_s = \bar{f}_s\, m_{s-1}$ (up to homotopy). Since $\bar{K}_s$ is an Eilenberg-MacLane object, it is sufficient to show that the induced cohomology maps are equal. Since $\{C_s\}$ is a resolution, $f^*_{s-1}: H^*(M_{s-1}) \longrightarrow H^*(K_{s-1})$ is a monomorphism and it thus suffices to show that $j^*_{s-1}\, f^*_s\, g^*_s = j^a_{s-1}\, m^a_{s-1} \bar{f}^*_s$. This follows from the assumption that $d\lambda_s = \lambda_{s-1} d$ (using the inductive hypothesis).

Given that $g_s f_s = \bar{f}_s m_{s-1}$ (up to homotopy), the whole of the diagram required by condition 3 follows by an obvious geometric argument.

I now **wish** to consider the convergence of this spectral sequence. By recalling the theory of exact couples, one writes down a portion of the rth derived couple. ($q^r_* =$ r-fold iteration of the map q_*.)

$$\boxed{\operatorname{Im} q^r_* : \pi_{t+r}(X, M_{s+r}) \longrightarrow \pi_t(X, M_s)}$$

$\uparrow$ J (defined by j_* into the second group)

$$\boxed{E^{s,t}_{r+1}: \text{ a sub quotient of } \pi_t(X, K_s)}$$

$\uparrow$ F (defined by f_* from the first group)

$$\boxed{\operatorname{Im} q^r_* : \pi_t(X, M_{s-1}) \longrightarrow \pi_{t-r}(X, M_{s-r-1})}$$

$\uparrow$ Q (defined by q_*)

$$\boxed{\operatorname{Im} q^r_* : \pi_{t+1}(X, M_s) \longrightarrow \pi_{t-r+1}(X, M_{s-r})}$$

If r is large compared with s, then the range and domain of Q can be identified with subgroups of $\pi_{t-s}(X, M_{-1}) = \pi_{t-s}(X, Y)$; the subgroups give the filtration defined by the images of $\pi_t(X, M_s)$.

For convergence, then, it remains only to prove the following lemma.

<u>Lemma 2</u>. $\bigcap_r \operatorname{Im} q^r : \pi_{t+r}(X, M_{s+r}) \longrightarrow \pi_t(X, M_s)$

consists of elements of odd order.

Since the spectral sequence is an invariant, it is sufficient to do this for a favorably chosen resolution. At this point I recall the hypothesis that $H^*(Y)$ is free over the exterior algebra E generated by Sq^1. We prove below that this allows us to find a resolution such that $Z_{s,t} = 0$ for $t < 2s + 2$. Hence $H_t(M_s; Z_2) = 0$ for $t < 2s + 2$. and (by Serre's mod C theorems)

$\pi_t(S^0, M_s)$ is an odd torsion group for $t < 2s + 2$

Hence $\pi_t(VS^0_i\, ; M_s)$ is an odd torsion group for $t < 2s + 2$ where VS^0_i is a one point union of copies of S^0. If X is finite dimensional, we deduce by exact sequence arguments that

$\pi_t(X, M_s)$ is an odd torsion group for $t < 2s + 2 - c$,

where c is a constant depending on X. Therefore $\pi_{t+r}(X, M_{s+r})$ is an odd torsion group for $t + r < 2s + 2r + 2 - c$; for a given s and t, this is true for sufficiently large r.

<u>Lemma 3</u>. Existence of a "nice" resolution.

<u>Remark 1</u>. M is free over E if and only if the homology of the module M with respect to Sq^1 as boundary operator is 0.

<u>Remark 2</u>. In an exact sequence

$$0 \longrightarrow M' \longrightarrow M \longrightarrow M'' \longrightarrow 0,$$

if two of the terms are free over E, then the third is also.

<u>Proof</u>: Remark 1 and the exact homology sequence.

<u>Remark 3</u>. If M is free over E and $0 \longleftarrow M \xleftarrow{\epsilon} C_o \longleftarrow \dots \longleftarrow C_s \longleftarrow \dots$ is a resolution, then each Z_s is free over E.

<u>Proof</u>: Follows by induction on s, applying remark 2 to the sequence $0 \longleftarrow Z_{s-1} \longleftarrow C_s \longleftarrow Z_s \longleftarrow 0$ and the fact that C_s is free over A which qua left module over E is itself a free E module.

<u>Remark 4</u>. We can choose C_s inductively so that $Z_{s,t} = 0$ for $t < 2s + 2$.

<u>Proof</u>: Suppose $C_o, \dots, C_{s-1}$ chosen such that $Z_{s-1,t} = 0$ for $t < 2s$. Choose E-free generators $q_1, \dots, q_n$ in $Z_{s-1,2s}$. Take corresponding A-free generators $h_1, \dots h_n$ in $C_{s,2s}$. Since $Sq^1 q_1, \dots, Sq^1 q_n$

are linearly independent, we have introduced no cycles in $C_{s,2s+1}$. Introduce no more A-free generators in $C_{s,2s+1}$ than are needed to map onto the remaining E free generators in $Z_{s-1,2s+1}$. We therefore have $Z_{s,t} = 0$ for $t < 2s + 2$ whatever is done in higher dimensions.

I want next to consider products in the spectral sequence. In the E_2 term of the special case $X = Y = S^0$ we have the cup products of homological algebra for $E_2^{s,t} = H^{s,t}(A) = \mathrm{Ext}_A^{s,t}(Z_2,Z_2)$. We also have products in ${}_2\pi_r(S^0,S^0)$; the product structure is given geometrically by the composition of maps. It is a theorem that one can introduce products into the whole of the spectral sequence, compatible with these two products in E_2 and E_∞, and so that d_r is a derivation, of course. This result is already in my paper in the Commentarii Mathematici Helvetici [7]. However, the result should be somewhat more general. Consider three stable complexes X, Y, Z, so that we have three spectral sequences

$$\mathrm{Ext}^{s,t}(H^*(Z),H^*(Y)) \underset{s}{\Longrightarrow} {}_2\pi_*(Y,Z)$$

$$\mathrm{Ext}^{s,t}(H^*(Y),H^*(X)) \underset{s}{\Longrightarrow} {}_2\pi_*(X,Y)$$

$$\mathrm{Ext}^{s,t}(H^*(Z),H^*(X)) \underset{s}{\Longrightarrow} {}_2\pi_*(X,Z)$$

One would hope that there would be a pairing which pairs the first two spectral sequences to the third, and which is compatible with the composition product.

$$_2\pi(Y, Z) \otimes {}_2\pi_*(X, Y) \longrightarrow {}_2\pi_*(X, Z)$$

and the cup-product in homological algebra. I have never written out any details for this generalized case, but I believe that Puppe, in Chicago, is preparing a paper which will include this.

My next topic is cohomology operations of the nth kind. I will be rather brief because it is not yet certain how far it is necessary to develop the theory. For example, for some purposes, you can take the spectral sequence that I have already developed, and use the differentials d_n as a substitute for cohomology operations of the nth kind. It is probably more satisfying to be somewhat more general. We have universal examples for primary operations, namely Eilenberg-MacLane objects. It is natural to see what we can construct using as our universal examples n-fold extensions of Eilenberg-MacLane objects. The corresponding notion in the category of spaces would be an iterated fibering. We would begin by taking a map $f_1: K_o \longrightarrow K_1$ of Eilenberg-MacLane spaces, and then construct the induced fibering. In the stable category we take a map $f_1: K_o \longrightarrow K_1$ and construct

$$K_0 \xrightarrow{f_1} K_1 \xrightarrow{\rho_1} M_1 = K_1 \underset{f_1}{\smile} CK_0 \xrightarrow{q_1} SK_0 \longrightarrow \cdots$$

Now we take K_2 and construct

$$M_1 \xrightarrow{f_2} K_2 \xrightarrow{\rho_2} M_2 \xrightarrow{q_2} SM_1$$

$$\vdots$$

$$M_{n-1} \xrightarrow{f_n} K_n \xrightarrow{\rho_n} M_n \xrightarrow{q_n} SM_{n-1}$$

If we look at this, we see that it is the same sort of thing we previously called a "realization," because we kept a little spare generality in hand for this purpose. (Strictly, to fix up the details, I have to define M_0 to be K_0 so that M_{-1} is a "point"; I also have K_s to be a "point" for $s > n$.) We can look at the spectral sequence of this realization. The first term is

$$\begin{array}{c} \mathrm{Hom}_A^*(C_0, H^*(X)) \\ \downarrow d^* \\ \mathrm{Hom}_A^*(C_1, H^*(X)) \\ \downarrow d^* \\ \vdots \\ \downarrow \\ \mathrm{Hom}_A^*(C_n, H^*(X)) \end{array}$$

where I have written C_s for $H^*(K_s)$. Now suppose that C is an A-face module on generators c_i $(i = 1, \ldots, m)$: Then an A map

$f: C \longrightarrow H^*(X)$ is determined by giving the elements $f(c_i) \in H^*(X)$. With this interpretation, each homomorphism d^* may be interpreted as a primary operation, from m variables to m' variables. Consequently each d_2 may be regarded as a function from the kernel of one primary operation to the cokernel of another primary operation. I am going to offer you the differential d_n as defining an operation of the nth kind, and I ought to verify that this agrees with one's notion of the usual procedure with universal examples. Our procedure is given by the following diagram:

$$\begin{array}{ccc} \pi_{t+n-1}(X, M_{n-1}) & \xrightarrow{(q_*)^{n-1}} & \boxed{\text{subgroup of } \pi_t(X, M_o)} \\ \downarrow F = f_* & & \nearrow J = \text{identity} \\ \pi_{t+n-1}(X, K_n) & \qquad E_n^{o,t} & (\text{subgroup of } \pi_t(X, K_o)) \end{array}$$

That is, you realize a cohomology m-tuple by a map from S^tX into K_o: you lift this, if you can, to a map, μ, into the universal example, M_{n-1}: you now regard $f^*: C_n \to H^*(M_{n-1})$ as giving you a m'-tuple in $H^*(M_{n-1})$ and you take the image of this m'-tuple by μ^* in $H^*(X)$. This is precisely one's ordinary notion of the procedure for defining an operation by means of a universal example. One comment is called for; I have supposed given the realization consisting of the M_s's and the K_s's. This supposition involves an irreducible element of geometry; for $n \geq 3$, not every chain complex

$$C_n \xrightarrow{d} C_{n-1} \longrightarrow C_1 \longrightarrow C_o$$

can be realized by K_s's and M_s's. C.R.F. Maunder has developed the theory in this direction. He has defined axiomatically the notion that an operation, Φ, is associated with a chain complex $C_n \rightarrow \ldots \rightarrow C_o$. He proves, for example, that if Φ is associated with $C_r \rightarrow \ldots \rightarrow C_1 \rightarrow C_o$ and Ψ is associated with $C_s' \rightarrow \ldots \rightarrow C_1' \rightarrow C_o'$, and if $C_o = C_s'$, then $\Phi\Psi$ is associated with

$$C_r \rightarrow \ldots \rightarrow C_1 \nearrow \overset{C_o = C_s'}{\longrightarrow} \searrow C_{s-1}' \rightarrow \ldots \rightarrow C_o' .$$

Similarly, he shows that the Spanier-Whitehead dual, $c\Phi$, of Φ is associated with a chain complex $cC_o \rightarrow cC_1 \rightarrow \ldots \rightarrow cC_r$ constructed from $C_r \rightarrow \ldots \rightarrow C_o$ by a well defined algebraic process.

Appendix to Lecture 4

The following table gives a Z_2 basis for $H^{s,t}(A)$ in the range of (s,t) indicated. The following differentials are known:

$$d_2 h_4 = h_0 h_3^2$$

$$d_3(h_0 h_4) = h_0 g$$

$$d_3(h_0^2 h_4) = h_0^2 g$$

$$d_3(h_2 h_4) = h_2 g$$

$$d_3(h_0 h_2 h_4) = h_0 h_2 g$$

$$d_3(h_0^2 h_2 h_4) = h_0^2 h_2 g.$$

The notation Px implies that this element corresponds under a periodicity isomporphism to the element x.

11	h_0^{11}									
10	h_0^{10}									
9	h_0^{9}									
8	h_0^{8}									
7	h_0^{7}									
6	h_0^{6}									
5	h_0^{5}									$Ph_1 = \langle h_3, h_0^4, h_1\rangle$
4	h_0^{4}							$h_0^3h_3$		$\langle h_2^2, h_0, h_1^2\rangle$
3	h_0^{3}			$h_1^3 = h_0^2h_2$				$h_0^2h_3$	$\langle h_2^2, h_0, h_1\rangle$	$h_2^3 = h_1^2h_3$
2	h_0^{2}		h_1^2	h_0h_2			h_2^2	h_0h_3	h_1h_3	
1	h_0	h_1		h_2				h_3		
s=0	1									

11								
10								
9								P^2h_1
8						$h_0^7h_4$		$P\langle h_2^2,h_0,h_1^2\rangle$
7		$\langle h_3,h_0^4,h_1^3\rangle$ $=Ph_1^3$				$h_0^6h_4$	$P\langle h_2^2,h_0,h_1\rangle$	h_1^3g $=P(h_1^2h_3)$
6	$Ph_1^2 =$ $\langle h_3,h_0^4,h_1^2\rangle$	$\langle h_3,h_0^4,h_0h_2\rangle$ $=Ph_0h_2$			h_0^2g	$h_0^5h_4$	$h_1^2g =$ $P(h_1h_3)$	h_0h_2g
5		$\langle h_3,h_0^4,h_2\rangle$ $=Ph_2$			h_0g	$h_0^4h_4;h_1g$		h_2g
4					$g =$ $\langle h_0,h_2^2,h_2^2,h_0\rangle$	$h_0^3h_4$		k
3					$h_0h_3^2$	$h_0^2h_4$		$h_1^2h_4$
2					h_3^2	h_0h_4	h_1h_4	
1						h_4		
s=0								
	10	11	12	13	14	15	16	17

5) Theorems of periodicity and approximation in homological algebra.

Let us begin by contrasting the spectral sequence I have developed with the classical method of killing homotopy groups, as applied to the calculation of stable homotopy group Both depend on a knowledge of the stable Eilenberg-MacLane groups $H^{n+q}(\pi, n; G)$ $(n > q)$ for some π and G. Neither of them is an algorithm. By an algorithm I would mean a procedure that comes provided with a guarantee that you can always compute any required group by doing a finite amount of work following the instructions blindly. In the case of the method of killing homotopy groups, you have no idea how far you can get before you run up against some ambiguity and don't know how to settle it. In the case of the spectral sequence, the situation is clearer: the groups $\mathrm{Ext}_A^{s,t}(H^*(Y), H^*(X))$ are recursively computable up to any given dimension; what is left to one's intelligence is finding the differentials in the spectral sequence, and the group extensions at the end of it.

This account would be perfectly satisfying to a mathematical logician: an algorithm is given for computing $\mathrm{Ext}_A^{s,t}(H^*(Y), H^*(X))$; none is given for computing d_r. The practical mathematician, however, is forced to admit that the intelligence of mathematicians is an asset at least as reliable as their willingness to do large amounts of tedio mechanical work. The history of the subject shows, in fact, that whenever a chance has arisen to show that a differential d_r is non-zero, the experts have fallen on it with shouts of joy - "Here is an interesting phenomenon! Here is a chance to do some nice, clean research!" - and they have solved the problem in short order. On the other hand, the calculation of $\mathrm{Ext}_A^{s,t}$ groups is necessary not only for this spectral sequence, but also for the study of cohomology operations of the nth kind: each such group can be calculated by a large amount of tedious mechanical work: but the process finds few people willing to take it on.

In this situation, what we want is theorems which tell us the value of the $\mathrm{Ext}_A^{s,t}$ groups. Now it is a fact that the $\mathrm{Ext}_A^{s,t}$ groups enjoy a certain limited amount of periodicity, and I would like to approach this topic in historical order.

First recall that last time I wrote down a basis for $\mathrm{Ext}_A^{s,t}\ (Z_2, Z_2)$ for small s and t:

4	h_0^4							$h_0^3h_3$	
3	h_0^3			$h_0^2h_2$				$h_0^2h_3$	
2	h_0^2		h_1^2	h_0h_2			h_2^2	h_0h_3	
1	h_0	h_1		h_2				h_3	
0	1								
	0	1	2	3	4	5	6	7	

$t - s \longrightarrow$

It was implied that $\mathrm{Ext}_A^{s,t}(Z_2, Z_2) = 0$ for larger values of s in the range $0 < t - s \leq 7$. This is actually a theorem, which is proved in [7].

<u>Theorem 1</u>. There is a numerical function, $f(s)$, such that:

(i) $\mathrm{Ext}_A^{s,t}(Z_2, Z_2) = 0$ for $s < t < f(s)$

(ii) $f(s) \geq 2s$

(iii) $f(s + s') \geq f(s) + f(s')$

(iv) $f(0) = 0$, $f(1) = 2$, $f(2) = 4$, $f(3) = 6$, $f(4) = 11$

The published proof of this theorem is by induction, and the induction involves $\mathrm{Ext}_A^{s,t}(M, Z_2)$ for A-modules M other than Z_2. We consider the exterior algebra E generated by Sq^1, so that we have an injection

$$i: E \longrightarrow A .$$

This induces

$$i^*: \mathrm{Ext}_A^{s,t}(M, Z_2) \longrightarrow \mathrm{Ext}_E^{s,t}(M, Z_2)$$

(For example, if $M = Z_2$, then $\mathrm{Ext}_E^{s,t}(Z_2, Z_2)$ is a polynomial algebra with h_0 as its generator. In general, if M is a module over E, then $Sq^1: M \longrightarrow M$ is a boundary operator on M and $\mathrm{Ext}_E^{s,t}(M, Z_2) \cong H^{t-s}(M)$ for $s > 0$, where H^* denotes the homology with respect to Sq^1).

What one proves, then is the following.

<u>Theorem 2</u>. Suppose $M_t = 0$ for $t < m$; with the same function $f(s)$ as in theorem 1, the map

$$i^*: \ \mathrm{Ext}_A^{s,t}(M, Z_2) \longrightarrow \mathrm{Ext}_E^{s,t}(M; Z_2)$$

is an isomorphism for $t < m + f(s)$.

In the same paper, I also conjecture that for $s = 2^n (n \geq 2)$ the best possible value of $f(s)$ is $f(2^n) = 3 \cdot 2^n - 1$. This conjecture is actually true. As a matter of fact, some correspondence with Liulevicius involved me in extended calculations which strongly suggested that the best possible function $f(s)$ is given by

$$f(4n) = 12n - 1 \qquad (\text{for } n > 0)$$

$$f(4n + 1) = 12n + 2$$

$$f(4n + 2) = 12n + 4$$

$$f(4n + 3) = 12n + 6.$$

This is actually true, so that the function $f(s)$ which gives the "edge" of the E_2 diagram is periodic with period 4 in s and with period 12 in t. The period in $t - s$ is therefore 8, and this strongly reminds us of Bott's results.

As a matter of fact more is true. Not only is the "edge" of the E_2 diagram periodic, but the groups near the edge are periodic: i.e. in a neighborhood N_0 of the line $t = 3s$, we have $H^{s,t}(A) \cong H^{s+4,t+12}(A)$.

More still is true. In a bigger neighborhood, N_k,

of the line $t = 3s$, the groups $H^{s,t}(A)$ are periodic with period $4 \cdot 2^k$ in s, $12 \cdot 2^k$ in t. The union of these neighborhoods, N_k, is the area $t < g(s)$ where $4s \leq g(s) \leq 6s$. (Possibly $g(s) = 2f(s)$, but I cannot give the exact value until I have refined my methods a little.)

Again, these periodicity theorems should not be restricted to the case of $\mathrm{Ext}_A^{s,t}(Z_2, Z_2)$. We should deal with $\mathrm{Ext}_A^{s,t}(M, Z_2)$. We deal with the case in which M is free over E, the exterior algebra generated by Sq^1: Theorem 2 shows that this is indispensable in the general case. Although this condition is not satisfied by the module Z_2, periodicity results for $\mathrm{Ext}_A^{s,t}(Z_2, Z_2)$ can be deduced from the following formula.

$$\mathrm{Ext}_A^{s,t}(Z_2, Z_2) \cong \mathrm{Ext}_E^{s,t}(Z_2, Z_2) + \mathrm{Ext}_A^{s-1,t}(I(A)/A\, Sq^1, Z_2)$$

Here $I(A)/A\, Sq^1$ is a free left module over E.

Well, now, let us see some details. In what follows, A_r will denote the algebra generated by $Sq^1, Sq^2, \ldots, Sq^{2^r}$ when r is finite; A_∞ will denote A. Note that $A_o = E$. For our first results, we assume that L is a left module over A_r, that L is free qua left module over A_o, and that $L_t = 0$ for $t < \ell$.

<u>Theorem 3.</u> (Vanishing). $\mathrm{Tor}_{s,t}^{A_r}(Z_2, L)$ and $\mathrm{Ext}_{A_r}^{s,t}(L, Z_2)$ are zero if $t < \ell + T(s)$ where T is the

numerical function defined by

$$T(4k) = 12k$$

$$T(4k+1) = 12k+2$$

$$T(4k+2) = 12k+4$$

$$T(4k+3) = 12k+7$$

<u>Theorem 4</u>. (Approximation). The maps

$$i_*: \mathrm{Tor}^{A\rho}_{s,t}(Z_2, L) \longrightarrow \mathrm{Tor}^{Ar}_{s,t}(Z_2, L)$$

and $i^*: \mathrm{Ext}^{s,t}_{A\rho}(L, Z_2) \longleftarrow \mathrm{Ext}^{s,t}_{Ar}(L, Z_2)$ are isomorphisms

if $0 < \rho < r$, $s \geq 1$ and $t < \ell + T(s-1) + 2^{\rho+1}$.

I will not give complete proofs, but I will try to give some of the ideas.

a) It is not too laborious to compute Tor^B and Ext_B where B is a small subalgebra of A. For example, suppose we consider the case of Theorem 3 in which $r = 1$ (so $B = A_1$, a finite algebra generated by Sq^1 and Sq^2) and let $L = A_0$. Then we can make an explicit resolution of A_0 over A_1, and we can see that theorem 3 is true.

b) If theorem 3 is true in the special case $r = R$ (some fixed value) and $L = A_0$, then it is true for $r = R$ whatever L is.

In fact, if we are given theorem 3 for the A_r module A_0, then by exact sequences we can obtain theorem 3 for any A_r-module A_o, then by exact sequences we can obtain theorem 3 for any A_r-module which can be written as a finite extension of modules isomorphic to A_o. This is sufficient.

At this stage we have obtained theorem 3 for the case $r = 1$.

c) Theorem 4 tends to support theorem 3. In fact, if we know that

$$i_*: \mathrm{Tor}^{A_1}_{s,t}(Z_2, L) \longrightarrow \mathrm{Tor}^{A_r}_{s,t}(Z_2, L) \quad \text{is an isomorphism,}$$

and that $\mathrm{Tor}^{A_1}_{s,t}(Z_2, L) = 0$, then $\mathrm{Tor}^{A_r}_{s,t}(Z_2, L) = 0$.

d) Theorem 3 tends to support theorem 4. In fact, we consider the map $A_r \otimes_{A_\rho} L \longrightarrow L$ and define K to be its kernal, so that

$$0 \longrightarrow K \longrightarrow A_r \otimes_{A} L \longrightarrow L \longrightarrow 0$$

is an exact sequence. Then we have the following diagram

$$\begin{array}{ccccc} & & \mathrm{Tor}^{A_\rho}_{s,t}(Z_2, L) & & \\ & & \downarrow \cong & \searrow i_* & \\ \mathrm{Tor}^{A_r}_{s,t}(Z_2,K) \to & \mathrm{Tor}^{A_r}_{s,t}(Z_2, A_r \otimes_{A_\rho} L) & \to & \mathrm{Tor}^{A_r}_{s,t}(Z_2,L) & \to \mathrm{Tor}^{A_r}_{s-1,t}(Z_2, \end{array}$$

The vertical map is an isomorphism by a standard result on changing rings, which is in Cartan-Eilenberg [8] for the ungraded case. Also, if $L_t = 0$ for $t < \ell$, then $K_t = 0$ for $t < \ell + 2^{\rho+1}$. Hence theorem 3 implies that

$\mathrm{Tor}^{A_r}_{s,t}(Z_2,K)$ and $\mathrm{Tor}^{A_r}_{s-1,t}(Z_2,K)$ are zero for $t < \ell + 2^{\rho+1} + T(s-1)$. This implies that i_* is an isomorphism in the same range.

Of course, in order to apply theorem 3, it is necssary to prove that K is free over A_0, and this is one of the places where we rely on a firm grasp of the structure of A.

Given these ideas, it is possible to prove theorems 3 and 4 simultaneously by induction over the dimensions. The details are somewhat tricky, and I will not try to rehearse them here. The inference (d) goes smoothly enough; but in the inference (c), the conclusion of theorem 4 does not apply to the entire range of dimensions which we wish to consider. It is therefore necessary to preserve not only the conclusion of theorem 4 from a previous stage of induction, but also the ;method of proof used in (d).

<u>Theorem 5</u>. (Periodicity) There exists an element $\tilde{\omega}_r$ in $\mathrm{Ext}^{s,t}_{A_r}(Z_2,Z_2)$ for $S = 2^r$, $t = 3\cdot 2^r$ $(r \geq 2)$ with the following properties.

(i) The map $x \to x\tilde{\omega}_r$: $\mathrm{Ext}^{s,t}_{A_r}(L,Z_2) \to \mathrm{Ext}^{s+2^r,t+3\ 2^4}_{A_r}(L,Z_2)$ is an isomorphism when L is A_0-free and $t \leq U(s) + \ell$ and $U(s)$ is a numerical function such that $4s \leq U(s) \leq 6s$.

(ii) $\iota^*(\tilde{\omega}_r) = (\tilde{\omega}_{r-1})^2$ (This says that the periodicity maps for different r are compatible.)

(iii) In a certain range, where

$$i^*: \mathrm{Ext}_{A_r}^{s,t}(L,Z_2) \longleftarrow \mathrm{Ext}_A^{s,t}(L,Z_2)$$

is an isomorphism, the periodicity isomorphism on the left is transported by i^* to the Massey product operation $x \longrightarrow \langle x, h_0^{2^r}, h_{r+1}\rangle$ on the right.

Remarks. I will try to make this plausible starting from the end and working forwards.

The Massey product $\langle x,y,z\rangle$ is defined only when $xy = 0$ and $yz = 0$. The fact that $xh_0^{2^r}$ is zero when x lies in a suitable range is guaranteed by theorem 3. The fact that $h_0^{2^r} h_{r+1} = 0$ (for $4 \geq 2$) was previously known and was proved by introducing Steenrod squaring operations into $H^*(A)$! However, it can be deduced from theorems 3 and 4.

I have next to recall that $H^*(A)$ can be defined as the cohomology of a suitable ring of co-chains, by using the bar-construction. In fact, $h^{2^r} h_{r+1}$ is the cohomology class determined by

$$[\underbrace{\xi_1|\xi_1| \cdots |\xi_1}_{2^r\text{-times}}|\xi_1^{2^{r+1}}] .$$

Therefore we have a formula

$$\delta c = [\xi_1|\xi_1| \cdots |\xi_1|\xi_1^{2^{r+1}}] .$$

Now consider $i: A_r \to A$ and apply i^* to the above formula. We have $i^*(\xi_1^{2^{r+1}}) = 0$, whence $\delta(i^*c) = 0$ and i^*c defines a class, $\tilde{\omega}_r$ in $H^{s,t}(A_r)$ for $s = 2^r$, $t = 3\cdot 2^r$. One checks that $\tilde{\omega}_r$ has the property (iii). $\tilde{\omega}_r$ is actually well defined by the above description.

We now begin an argument like the former one.

Step (a). The homomorphism

$$x \to x(i^*\tilde{\omega}_2) : \mathrm{Ext}_{A_1}^{s,t}(A_0, Z_2 \to \mathrm{Ext}_{A_1}^{s+4,t+12}(A_0, Z_2)$$

is an isomorphism for $s \geq 0$.

Proof by explicit computation.

Step (b). The homomorphism

$$x \to x(i^*\tilde{\omega}_2) : \mathrm{Ext}_{A_1}^{s,t}(L, Z_2) \to \mathrm{Ext}^{s+r,t+12}(L, Z_2)$$

is an isomorphism for $s > 0$ if L is A_0-free.

Proof: by taking successive extensions of A_1-modules isomorphic to A_0. (Since the homomorphism $x \to x(i^*\tilde{\omega}_2)$ is natural we can use arguments based on the Five Lemma.)

Step (c). It is now clear that $\mathrm{Ext}_{A_r}^{s,t}(L, Z_2)$ is periodic in the small range where it is isomorphic to $\mathrm{Ext}_{A_1}^{s,t}(L, Z_2)$. We now extend this result up the dimensions by induction. We form the exact sequence

$0 \longrightarrow K \longrightarrow A_r \otimes_{A_1} L \longrightarrow L \longrightarrow 0$. $\mathrm{Ext}_{A_r}(A_r \otimes_{A_1} L; Z_2) \cong$ $\mathrm{Ext}_{A_1}(L; Z_2)$ and this is periodic by step (b). Also, if $L_t = 0$ for $t < 1$, then $K_t = 0$ for $t < L+4$; so we can use the inductive hypothesis on K .

I remark that the reason this proof does not give the best value of the function $U(s)$ is that I started with calculations over A_2 , one could perhaps extract a best possible value for $U(s)$.

6) Comments on prospective applications of 5), work in progress, etc.

Once agin, I would like to hang out a large sign saying "Provisional--Work in Progress." My first remark however is a theorem.

Remark 1. The theorems of the previous lecture allow one to put an explicit upper bound on the order of elements in ${}_2\pi_r(S^0,S^0)$. In fact, we have filtered ${}_2\pi_r(S^0,S^0)$ so that the composition quotients are vector spaces over Z_2, and we have put explicit upper bounds on the length of the composition series. For large r, the bound on the order of elements is appriximately $2^{(1/2\ r)}$; the previous best value, due to I. M. James, was approximately 2^r for large r.

Question 1. I've remarked that as soon as you define new cohomology operations you are entitled to some dividiend in the way of calculation and results. Stable cohomology operations of the nth kind are associated with free chain complexes over A. The work of the last lecture leads one to consider a lot of chain complexes over A which are periodic; the fundamental one is

$$\ldots \to A \xrightarrow{x \to xSq^{0,1}} A \xrightarrow{x \to xSq^{0,1}} A \to \ldots \ .$$

In more familiar notation $Sq^{0,1} = Sq^3 + Sq^2Sq^1$.

One can certainly construct cohomology operations corresponding to the fundamental chain complex written down above; the proof relies on Botts' work. It is also possible to construct cohomology operations corresponding to a number of other periodic chain complexes; but the general situation is not clear.

Question 2. Behavior of the J-homomorphism. One may calculate the groups $\mathrm{Ext}_A^{s,t}(Z_2,Z_2)$ in a neighborhood of the line $t = 3s$, and it is plausible to conjecture that certain of these represent the image of the J-homomorphism in dimensions $8k$, $8k + 1$, $8k + 3$.

Question 3. Consider the spectral sequence $\mathrm{Ext}_A^{s,t}(Z_2,Z_2)$. Consider the differentials which arrive in a neighborhood of periodicity N_k, and originate (i) in a neighborhood of periodicity N_ℓ with $\ell \geq k$, or (ii) from the region of non-periodicity. Do these differentials show periodicity? (I think it is implausible to suppose that they show periodicity with as small a period as that which obtains in N_k.)

Let us make the question stronger. Can one find subgroups of ${}_2\pi_r(S^0,S^0)$ which display periodicity? The first periodicity operation should be

$$x \longrightarrow \langle x, 16\iota, \sigma \rangle$$

(where the bracket is a Toda bracket, and ι,σ are generators for the 0-stem and 7-stem). Further periodicity operations might be

$$x \longrightarrow \langle x, 2^f \iota, g_{k+1} \rangle$$

where g_{k+1} is an element of the greatest possible order in the $(2^{k+1} - 1)$ stem, and 2^f is the order of g_{k+1}.

The whole of this question is highly speculative. It is quite obscure how one could ever isolate the relevant sugroups directly, without bringing in the machinery introduced above.

Perhaps one can isolate the essence of questions 1 and 3 in a further question.

Question 4. What geometric phenomena can one find which show a periodicity and which on passing to algebra give the sort of periodicity encountered in the last lecture? The question is wide open.

Now I want to talk a little about the vector field question. It is classical that one can reduce this question to studying the homotopy theory of real projective spaces; let us recall how this is done.

One may define a map

$$RP^{n-1} \longrightarrow SO(n)$$

as follows. Choose a base point e in S^{n-1}. To each point $y \in S^{n-1}$, assign the following rotation: first reflect S^{n-1} in the hyperplane perpendicular to e, then reflect S^{n-1}

in the hyperplane perpendicular to y . Since y and its antipode give the same rotation, we have defined a map

$$RP^{n-1} \longrightarrow SO(n) .$$

By attention to detail, you can make the following diagram commutative.

$$\begin{array}{ccc}
RP^{n-1} & \longrightarrow & SO(n) \\
\downarrow & & \downarrow \\
RP^{n-1}/RP^{n-r-2} & \longrightarrow & SO(n)/SO(n-r-1) \\
\downarrow & & \downarrow \\
RP^{n-1}/RP^{n-2} = S^{n-1} & \xrightarrow{\text{degree } 1} & SO(n)/SO(n-1) = S^{n-1} .
\end{array}$$

It follows that if we can construct a lifting

$RP^{n-1}/RP^{n-r-2} \underset{\lambda}{\rightleftarrows} S^{n-1}$, then the required r-field exists.

Conversely, if the dimensions n and r are favourably disposed, then RP^{n-1}/RP^{n-r-2} is equivalent to $SO(n)/SO(n-r-1)$ up to the required dimension, so that the existence of the lifting λ is necessary and sufficient for the existence of an r-field on S^{n-1} . We have to decide, therefore, whether the top-dimensional homology class in RP^{n-1}/RP^{n-r-2} spherical. It is sufficient to show that it is not spherical after suspension.

Let us examine the spectral sequence

$$\mathrm{Ext}_A^{*,*}(H^*(RP^{n-1}/RP^{n-r-2}),Z_2) \Longrightarrow {}_2\pi_*^S(RP^{n-1}/RP^{n-r-2}) .$$

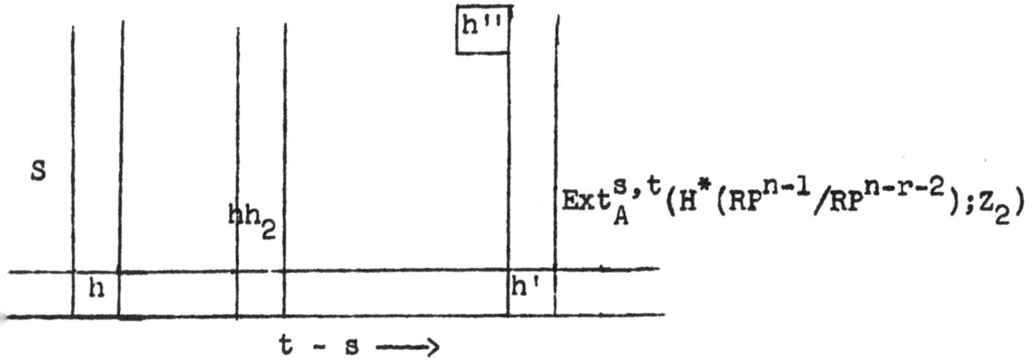

Let us assume that we have $n = 2^m$, $r = 8k+4$. Then the top and bottom cohomology classes in RP^{n-1}/RP^{n-r-2} correspond to x^{2^m-1}, x^{2^m-8k-5} in RP^∞ . These correspond to generators h,h' in Ext^0 . Let us calculate $Sq^4x^{2^m-8k-5}$: whether this is zero or not depends only on the congruence class mod 8 of $2^m-8k - 5$, and $Sq^4x^3 = 0$, so $Sq^4x^{2^m-8h-5} = 0$. This gives a class hh_2 in $Ext^{s,t}$ for $s = 1$, $t-s = 2^m - 8h - 2$. By periodicity we get a class h'' in $Ext^{s,t}$ for $S = 4k + 1$, $t - s = 2^m - 2$.

Question 5. Is $d_rh' = 0$ for $r < 4k + 1$? Probably the answer is yes. If the answer is no, then h' is not spherical anyway, so we don't need to worry about this question.

Question 6. Is $h'' = d_rx$ for $r < 4k +1$? One hopes that the answer is no.

Question 7. Is $d_{4k+1}k' = h''$? One hopes that the answer is yes.

Bibliography

[1] J. Adem. The iteration of Steenrod Squares in algebraic topology, Proc. Nat. Acad. Sci. 38 (1952), 720-726.

[2] J. P. Serre. Cohomologie modulo 2 des complexes d'Eilenberg-MacLane, Comment. Math. Helv. 27 (1953), 198-231.

[3] S. Eilenberg. Cohomology and continuous mappings, Annals of Math. 41 (1940) 231-251.

[4] J. C. Moore. Seminaire H. Cartan 1954-55, Algèbres d'Eilenberg-MacLane et Homotopie; Exposés 2 et 3.

[5] E. C. Zeeman. A proof of the comparison theorem for spectral sequences, Proc. Camb. Phil. Soc. 53 (1957), 57-62.

[6] J. Milnor. The Steenrod Algebra and its dual, Annals of Mathematics 67 (1958), 150-171.

[7] J. F. Adams. On the structure and applications of the Steenrod algebra, Comment. Math. Helvet. 32 (1958), 180-214.

[8] Cartan-Eilenberg. Homological Algebra, Princeton University Press, 1956, Chap. vi, sec. 4, pp 116-119.

Appendix

The following appendix is organised as a series of notes on the main text. Its object is to mention more recent developments in the topics studied, and also to correct certain errata in the original text.

P 2, line 4, Conjecture. This conjecture is now proved for k odd. For k even it is known that $J(\pi_{4k-1}(SO))$ is either Z_m or Z_{2m}. See [2], p 147. Theorem 3.7.

P 2, line 7, Conjecture. This conjecture is now proved. See [4]. Theorems 1.1 and 1.3; alternatively, combine [2], Example 3.5 with [3], Theorem 1.1.

P 3, line 1, Conjecture. This conjecture is now proved. See [1], Theorem 1.1.

P 3, lines 4-13. These comments are therefore out of date.

P 5, line 6, and later. For "Adam", read "Adem".

P19, last line. For "may", read "map".

P21, next-to-last line. For "(ab)", read "φ^*(ab)".

P23, last 4 lines and P24, first 9 lines. It turns out that this "example" (the Spanier-Whitehead dual of an Eilenberg - Mac Lane object) is ill chosen. Many "hares" would reject it, and many quite idealistic systems of stable homotopy theory will not include it.

P 24, lines 11-14. Let us start from some category of CW- complexes and cellular maps. Then the requisite constructions are: -

(i) To take a direct limit of categories under the suspension functor. This is essentially the way the Spanier-Whitehead category is obtained from the category of finite CW-complexes and homotopy classes of maps.

(ii) To take a direct limit of objects under inclusion maps. (This is essentially the way one gets an infinite CW-complex from its finite subcomplexes). Similarly for maps.

(iii) To leave the objects of the category alone, but pass from maps to homotopy classes of maps.

I first heard these ideas suggested by J. W. Milnor in private conversations. For details, see [5].

P 28, line 4. Delete the word " one ".

P 31, line 3 from foot. For "end" read "ends".

P 32, line 5 from foot. For " $Y \; Y$ ", read " $Y \vee Y$ ".

P 34, line 9. Insert a cup sign where needed.

P 35, line 7. For "show", read "shows".

P 42, line 7. The spectral sequence can be made to work with integral coefficients.

P 43, line 1. For " 2-1 ", read " s-1".

P 55, lines 8 - 10. Withdraw the assertions concerning $d_3 (h_2 h_4)$, $d_3 (h_0 h_2 h_4)$ and $d_3 (h_0^2 h_2 h_4)$.

P 58, line 5 from foot. For "dr", read "d_r".

P 63, line 10 to p 65, line 9 from foot. It is also possible to separate the proofs of the Vanishing Theorem and the Approximation Theorem, as follows.

(a) One first proves by computation that the Vanishing Theorem is true in the special case $r = \infty$, $L = A_0$, $s \leq 4$.

(b) By exact sequences we generalise this to the case $r = \infty$, $s \leq 4$, any L.

(c) By considering the first part of a minimal resolution

$$0 \leftarrow L \xleftarrow{\varepsilon} C_0 \leftarrow C_1 \leftarrow C_2 \leftarrow C_3 \leftarrow M \leftarrow 0$$

and applying an inductive hypothesis to M, we now generalise this to the case $r = \infty$, any s.

(d) We now generalise this to any r by change-of-rings.

(e) We now deduce the Approximation Theorem as on p 64 d).

P 63, last line and p 64, first line. Omit the words which have been repeated, e.g. from "then by exact sequences" on p. 63 to " A_r - module A_0 " on p 64 (inclusive).

P 64, line 12 and later. For "Ar", "Aρ", read "A_r", "A_ρ".

P 64, line 13. For "kernal", read "kernel".

P 68. Between lines 6 and 7, insert the following omitted words : " calculations over A_1; by starting with ".

P 70, line 7, Question 2. The conjecture remains plausible.

P 70, Question 3 and p 71, Question 4. These questions remain open.

P 71, line 10. For " 1 ", read " 2 ".

P 71, line 9 from foot to foot of p 73. In view of the work in [1] , this discussion is completely out of date, and can at most have historical interest as having lead to the methods of [1] ; see the comments in [1], last paragraph of p. 604 and first paragraph of p. 605.

References.

[1] J.F. Adams, Vector Fields on Spheres, Annals of Math. 75 (1962) pp 603-632.

[2] ________, On the Groups J(X) - II, Topology 3 (1965) pp 137-171.

[3] ________, On the Groups J(X) - III, Topology 3 (1965) pp 193-222.

[4] ________, On the Groups J(X) - IV, Topology, to appear.

[5] J.M. Boardman, Thesis, Cambridge 1964.

Offsetdruck: Julius Beltz, Weinheim/Bergstr.

ecture Notes in Mathematics

ier erschienen/Already published

1: J. Wermer, Seminar über Funktionen-Algebren. IV, 30 Seiten. 4. DM 3,80 / $ 0.95

2: A. Borel, Cohomologie des espaces localement compacts rès. J. Leray. IV, 93 pages. 1964. DM 9,– / $ 2.25

3: J. F. Adams, Stable Homotopy Theory. Third edition. IV, 78 pages. 9. DM 8,– / $ 2.00

4: M. Arkowitz and C. R. Curjel, Groups of Homotopy Classes. revised edition. IV, 36 pages. 1967. DM 4,80 / $ 1.20

5: J.-P. Serre, Cohomologie Galoisienne. Troisième édition. 214 pages. 1965. DM 18,– / $ 4.50

6: H. Hermes, Eine Termlogik mit Auswahloperator. IV, 42 Seiten. 5. DM 5,80 / $ 1.45

7: Ph. Tondeur, Introduction to Lie Groups and Transformation ups. VIII, 176 pages. 1965. DM 13,50 / $ 3.40

8: G. Fichera, Linear Elliptic Differential Systems and Eigenvalue lems. IV, 176 pages. 1965. DM 13,50 / $ 3.40

9: P. L. Ivănescu, Pseudo-Boolean Programming and Applications. 0 pages. 1965. DM 4,80 / $ 1.20

10: H. Lüneburg, Die Suzukigruppen und ihre Geometrien. VI, eiten. 1965. DM 8,– / $ 2.00

11: J.-P. Serre, Algèbre Locale. Multiplicités. Rédigé par P. Gabriel. onde édition. VIII, 192 pages. 1965. DM 12,– / $ 3.00

12: A. Dold, Halbexakte Homotopiefunktoren. II, 157 Seiten. 1966. 2,– / $ 3.00

13: E. Thomas, Seminar on Fiber Spaces. IV, 45 pages. 1966. ,80 / $ 1.20

14: H. Werner, Vorlesung über Approximationstheorie. IV, 184 Sei- nd 12 Seiten Anhang. 1966. DM 14,– / $ 3.50

15: F. Oort, Commutative Group Schemes. VI, 133 pages. 1966. ,80 / $ 2.45

16: J. Pfanzagl and W. Pierlo, Compact Systems of Sets. IV, ages. 1966. DM 5,80 / $ 1.45

17: C. Müller, Spherical Harmonics. IV, 46 pages. 1966. ,– / $ 1.25

18: H.-B. Brinkmann und D. Puppe, Kategorien und Funktoren. 07 Seiten, 1966. DM 8,– / $ 2.00

19: G. Stolzenberg, Volumes, Limits and Extensions of Analytic ties. IV, 45 pages. 1966. DM 5,40 / $ 1.35

20: R. Hartshorne, Residues and Duality. VIII, 423 pages. 1966. 0,– / $ 5.00

21: Seminar on Complex Multiplication. By A. Borel, S. Chowla, Herz, K. Iwasawa, J.-P. Serre. IV, 102 pages. 1966. DM 8,– / $ 2.00

22: H. Bauer, Harmonische Räume und ihre Potentialtheorie. 5 Seiten. 1966. DM 14,– / $ 3.50

3: P. L. Ivănescu and S. Rudeanu, Pseudo-Boolean Methods for ent Programming. 120 pages. 1966. DM 10,– / $ 2.50

24: J. Lambek, Completions of Categories. IV, 69 pages. 1966. ,80 / $ 1.70

5: R. Narasimhan, Introduction to the Theory of Analytic Spaces. 3 pages. 1966. DM 10,– / $ 2.50

26: P.-A. Meyer, Processus de Markov. IV, 190 pages. 1967. ,– / $ 3.75

27: H. P. Künzi und S. T. Tan, Lineare Optimierung großer me. VI, 121 Seiten. 1966. DM 12,– / $ 3.00

8: P. E. Conner and E. E. Floyd, The Relation of Cobordism to ories. VIII, 112 pages. 1966. DM 9,80 / $ 2.45

9: K. Chandrasekharan, Einführung in die Analytische Zahlen- e. VI, 199 Seiten. 1966. DM 16,80 / $ 4.20

0: A. Frölicher and W. Bucher, Calculus in Vector Spaces without X, 146 pages. 1966. DM 12,– / $ 3.00

1: Symposium on Probability Methods in Analysis. Chairman. Kappos. IV, 329 pages. 1967. DM 20,– / $ 5.00

2: M. André, Méthode Simpliciale en Algèbre Homologique et re Commutative. IV, 122 pages. 1967. DM 12,– / $ 3.00

33: G. I. Targonski, Seminar on Functional Operators and ons. IV, 110 pages. 1967. DM 10,– / $ 2.50

4: G. E. Bredon, Equivariant Cohomology Theories. VI, 64 pages. DM 6,80 / $ 1.70

5: N. P. Bhatia and G. P. Szegö, Dynamical Systems. Stability and Applications. VI, 416 pages. 1967. DM 24,– / $ 6.00

6: A. Borel, Topics in the Homology Theory of Fibre Bundles. pages. 1967. DM 9,– / $ 2.25

Vol. 37: R. B. Jensen, Modelle der Mengenlehre. X, 176 Seiten. 1967. DM 14,– / $ 3.50

Vol. 38: R. Berger, R. Kiehl, E. Kunz und H.-J. Nastold, Differentialrechnung in der analytischen Geometrie IV, 134 Seiten. 1967 DM 12,– / $ 3.00

Vol. 39: Séminaire de Probabilités I. II, 189 pages. 1967. DM 14,– / $ 3.50

Vol. 40: J. Tits, Tabellen zu den einfachen Lie Gruppen und ihren Darstellungen. VI, 53 Seiten. 1967. DM 6.80 / $ 1.70

Vol. 41: A. Grothendieck, Local Cohomology. VI, 106 pages. 1967. DM 10,– / $ 2.50

Vol. 42: J. F. Berglund and K. H. Hofmann, Compact Semitopological Semigroups and Weakly Almost Periodic Functions. VI, 160 pages. 1967. DM 12,– / $ 3.00

Vol. 43: D. G. Quillen, Homotopical Algebra. VI, 157 pages. 1967. DM 14,– / $ 3.50

Vol. 44: K. Urbanik, Lectures on Prediction Theory. IV, 50 pages. 1967. DM 5,80 / $ 1.45

Vol. 45: A. Wilansky, Topics in Functional Analysis. VI, 102 pages. 1967. DM 9,60 / $ 2.40

Vol. 46: P. E. Conner, Seminar on Periodic Maps. IV, 116 pages. 1967. DM 10,60 / $ 2.65

Vol. 47: Reports of the Midwest Category Seminar I. IV, 181 pages. 1967. DM 14,80 / $ 3.70

Vol. 48: G. de Rham, S. Maumary et M. A. Kervaire, Torsion et Type Simple d'Homotopie. IV, 101 pages. 1967. DM 9,60 / $ 2.40

Vol. 49: C. Faith, Lectures on Injective Modules and Quotient Rings. XVI, 140 pages. 1967. DM 12,80 / $ 3.20

Vol. 50: L. Zalcman, Analytic Capacity and Rational Approximation. VI, 155 pages. 1968. DM 13.20 / $ 3.40

Vol. 51: Séminaire de Probabilités II. IV, 199 pages. 1968. DM 14,– / $ 3.50

Vol. 52: D. J. Simms, Lie Groups and Quantum Mechanics. IV, 90 pages. 1968. DM 8,– / $ 2.00

Vol. 53: J. Cerf, Sur les difféomorphismes de la sphère de dimension trois (Γ_4= O). XII, 133 pages. 1968. DM 12,– / $ 3.00

Vol. 54: G. Shimura, Automorphic Functions and Number Theory. VI, 69 pages. 1968. DM 8,– / $ 2.00

Vol. 55: D. Gromoll, W. Klingenberg und W. Meyer, Riemannsche Geometrie im Großen. VI, 287 Seiten. 1968. DM 20,– / $ 5.00

Vol. 56: K. Floret und J. Wloka, Einführung in die Theorie der lokalkonvexen Räume. VIII, 194 Seiten. 1968. DM 16,– / $ 4.00

Vol. 57: F. Hirzebruch und K. H. Mayer, O (n)-Mannigfaltigkeiten, exotische Sphären und Singularitäten. IV, 132 Seiten. 1968. DM 10,80 / $ 2.70

Vol. 58: Kuramochi Boundaries of Riemann Surfaces. IV, 102 pages. 1968. DM 9,60 / $ 2.40

Vol. 59: K. Jänich, Differenzierbare G-Mannigfaltigkeiten. VI, 89 Seiten. 1968. DM 8,– / $ 2.00

Vol. 60: Seminar on Differential Equations and Dynamical Systems. Edited by G. S. Jones. VI, 106 pages. 1968. DM 9,60 / $ 2.40

Vol. 61: Reports of the Midwest Category Seminar II. IV, 91 pages. 1968. DM 9,60 / $ 2.40

Vol. 62: Harish-Chandra, Automorphic Forms on Semisimple Lie Groups X, 138 pages. 1968. DM 14,– / $ 3.50

Vol. 63: F. Albrecht, Topics in Control Theory. IV, 65 pages. 1968. DM 6,80 / $ 1.70

Vol. 64: H. Berens, Interpolationsmethoden zur Behandlung von Approximationsprozessen auf Banachräumen. VI, 90 Seiten. 1968. DM 8,– / $ 2.00

Vol. 65: D. Kölzow, Differentiation von Maßen. XII, 102 Seiten. 1968. DM 8,– / $ 2.00

Vol. 66: D. Ferus, Totale Absolutkrümmung in Differentialgeometrie und -topologie. VI, 85 Seiten. 1968. DM 8,– / $ 2.00

Vol. 67: F. Kamber and P. Tondeur, Flat Manifolds. IV, 53 pages. 1968. DM 5,80 / $ 1.45

Vol. 68: N. Boboc et P. Mustată, Espaces harmoniques associés aux opérateurs différentiels linéaires du second ordre de type elliptique. VI, 95 pages. 1968. DM 8,60 / $ 2.15

Vol. 69: Seminar über Potentialtheorie. Herausgegeben von H. Bauer. VI, 180 Seiten. 1968. DM 14,80 / $ 3.70

Vol. 70: Proceedings of the Summer School in Logic. Edited by M. H. Löb. IV, 331 pages. 1968. DM 20,– / $ 5.00

Vol. 71: Séminaire Pierre Lelong (Analyse), Année 1967 – 1968. VI, 19 pages. 1968. DM 14,– / $ 3.50

Bitte wenden / Continued

Vol. 72: The Syntax and Semantics of Infinitary Languages. Edited by J. Barwise. IV, 268 pages. 1968. DM 18,– / $ 4.50

Vol. 73: P. E. Conner, Lectures on the Action of a Finite Group. IV, 123 pages. 1968. DM 10,– / $ 2.50

Vol. 74: A. Fröhlich, Formal Groups. IV, 140 pages. 1968. DM 12,– / $ 3.00

Vol. 75: G. Lumer, Algèbres de fonctions et espaces de Hardy. VI, 80 pages. 1968. DM 8,– / $ 2.00

Vol. 76: R. G. Swan, Algebraic K-Theory. IV, 262 pages. 1968. DM 18,– / $ 4.50

Vol. 77: P.-A. Meyer, Processus de Markov: la frontière de Martin. IV, 123 pages. 1968. DM 10,– / $ 2.50

Vol. 78: H. Herrlich, Topologische Reflexionen und Coreflexionen. XVI, 166 Seiten. 1968. DM 12,– / $ 3.00

Vol. 79: A. Grothendieck, Catégories Cofibrées Additives et Complexe Cotangent Relatif. IV, 167 pages. 1968. DM 12,– / $ 3.00

Vol. 80: Seminar on Triples and Categorical Homology Theory. Edited by B. Eckmann. IV, 398 pages. 1969. DM 20,– / $ 5.00

Vol. 81: J.-P. Eckmann et M. Guenin, Méthodes Algébriques en Mécanique Statistique. VI, 131 pages. 1969. DM 12,– / $ 3.00

Vol. 82: J. Wloka, Grundräume und verallgemeinerte Funktionen. VIII, 131 Seiten. 1969. DM 12,– / $ 3.00

Vol. 83: O. Zariski, An Introduction to the Theory of Algebraic Surfaces. IV, 100 pages. 1969. DM 8,– / $ 2.00

Vol. 84: H. Lüneburg, Transitive Erweiterungen endlicher Permutationsgruppen. IV, 119 Seiten. 1969. DM 10,– / $ 2.50

Vol. 85: P. Cartier et D. Foata, Problèmes combinatoires de commutation et réarrangements. IV, 88 pages. 1969. DM 8,– / $ 2.00

Vol. 86: Category Theory, Homology Theory and their Applications I. Edited by P. Hilton. VI, 216 pages. 1969. DM 16,– / $ 4.00

Vol. 87: M. Tierney, Categorical Constructions in Stable Homotopy Theory. IV, 65 pages. 1969. DM 6,– / $ 1.50

Vol. 88: Séminaire de Probabilités III. IV, 229 pages. 1969. DM 18,– / $ 4.50

Vol. 89: Probability and Information Theory. Edited by M. Behara, K. Krickeberg and J. Wolfowitz. IV, 256 pages. 1969. DM 18,– / $ 4.50

Vol. 90: N. P. Bhatia and O. Hajek, Local Semi-Dynamical Systems. II, 157 pages. 1969. DM 14,– / $ 3.50

Vol. 91: N. N. Janenko, Die Zwischenschrittmethode zur Lösung mehrdimensionaler Probleme der mathematischen Physik. VIII, 194 Seiten. 1969. DM 16,80 / $ 4.20

Vol. 92: Category Theory, Homology Theory and their Applications II. Edited by P. Hilton. V, 308 pages. 1969. DM 20,– / $ 5.00

Vol. 93: K. R. Parthasarathy, Multipliers on Locally Compact Groups. III, 54 pages. 1969. DM 5,60 / $ 1.40

Vol. 95: A. S. Troelstra, Principles of Intuitionism. II, 111 pages. 1969. DM 10,– / $ 2.50

Vol. 97: S. O. Chase and M. E. Sweedler, Hopf Algebras and Galois theory. II, 133 pages. 1969. DM 10,– / $ 2.50

Vol. 98: M. Heins, Hardy Classes on Riemann Surfaces. III, 106 pages. 1969. DM 10,– / $ 2.50

Vol. 99: Category Theory, Homology Theory and their Applications III. Edited by P. Hilton. IV, 489 pages. 1969. DM 24,– / $ 6.00

Vol. 100: M. Artin and B. Mazur, Etale Homotopy. II, 169 Seiten. 1969. DM 12,– / $ 3.00

Vol. 101: G. P. Szegö et G. Treccani, Semigruppi di Trasformazioni Multivoche. VI, 177 pages. 1969. DM 14,– / $ 3.50

Vol. 102: F. Stummel, Rand- und Eigenwertaufgaben in Sobolewschen Räumen. VIII, 386 Seiten. 1969. DM 20,– / $ 5.00

Vol. 103: Lectures in Modern Analysis and Applications I. Edited by C. T. Taam. VII, 162 pages. 1969. DM 12,– / $ 3.00

Vol. 104: G. H. Pimbley, Jr., Eigenfunction Branches of Nonlinear Operators and their Bifurcations. II, 128 pages. 1969. DM 10,– / $ 2.50